Praise for *Baby Meets World*

"Nicholas Day has birthed a perfect book: expertly researched, beautifully written, wise, warm, honest, funny. What makes it fascinating is the same thing that makes it reassuring: When it comes to caring for babies, there has never been one right way to do it, but dozens—contradictory, bizarre (goat wet nurses!), hilarious in retrospect but always well-meant. If you have a baby, expect a baby, were or still are a baby, this is a book for you."

—Mary Roach, author of *Gulp*

"*Baby Meets World* is a breath of fresh air for parents increasingly pressured to do the next 'right' thing for their children. By exploring the wondrous complexities of early development in the context of personal experience as well as cultural norms, fads, and fancies, Nicholas Day provides a fascinating, entertaining, and ultimately reassuring look at what babies really need."

—Susan Linn, author of *The Case for Make Believe: Saving Play in a Commercialized World*

"Babies, yours or anyone else's, won't look the same after you've read *Baby Meets World*. Nicholas Day unearths the many peculiar things 'experts' and other adults have believed about infancy down the ages. With a fresh curiosity any baby could admire, he also pays rapt attention to some very normal things all babies do. He makes the familiar strange and utterly fascinating. What better antidote to all our fretting?"

—Ann Hulbert, author of *Raising America: Experts, Parents, and a Century of Advice About Children*

"With wry humor and sharp writing, Nicholas Day explains how— as in childbirth—raising a baby is often a reflection of the time and place in which it happens. Most important, he offers the perspective parents so often lack in the fog of battle. If you read one parenting book, make this it."

—Tina Cassidy, author of *Birth:*
The Surprising History of How We Are Born

"The challenges that face new parents are timeless, but as this wry, accessible, and provocative book demonstrates, the responses of various cultures and historical eras differ profoundly. Anyone interested in infancy, or simply the significance of a baby's smile will benefit greatly from this book's astute insights drawn from history, anthropology, physiology, and developmental psychology."

—Steven Mintz, author of *Huck's Raft:*
A History of American Childhood

"In *Baby Meets World*, Nicholas Day brings us the book about infants that parents should have had all along: smart, funny, and tender, with just the right amount of edge. Engaging and enlightening, it belongs in the diaper bag of every new mom and dad."

—Annie Murphy Paul, author of *Origins:*
How the Nine Months Before Birth Shape the Rest of Our Lives

Baby Meets World:
Suck, Smile, Touch, Toddle

A Journey Through Infancy

Nicholas Day

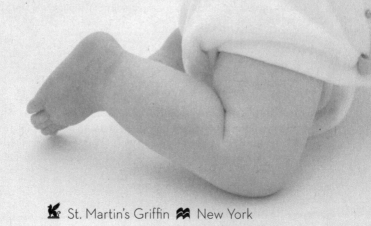

St. Martin's Griffin ⚏ New York

For my parents

BABY MEETS WORLD. Copyright © 2013 by Nicholas Day. All rights reserved. Printed in the United States of America. For information, address St. Martin's Press, 175 Fifth Avenue, New York, N.Y. 10010.

www.stmartins.com

Designed by Phil Mazzone

The Library of Congress has cataloged the hardcover edition as follows:

Day, Nicholas.
    Baby meets world: suck, smile, touch, toddle / Nicholas Day. — First edition.
        p. cm.
    ISBN 978-0-312-59134-2 (hardcover)
    ISBN 978-1-250-03861-6 (e-book)
    1. Infants—Development. 2. Infant psychology. 3. Parent and infant. I. Title.
    HQ774.D39 2013
    155.42'2—dc23

                                                                    2013002620

ISBN 978-1-250-04481-5 (trade paperback)

St. Martin's Griffin books may be purchased for educational, business, or promotional use. For information on bulk purchases, please contact Macmillan Corporate and Premium Sales Department at 1-800-221-7945, extension 5442, or write specialmarkets@macmillan.com.

First St. Martin's Griffin Edition: April 2014

10 9 8 7 6 5 4 3 2 1

Contents

Baby Meets World

Introduction

There is nothing less permanent in life than infancy: the point of being a baby is to stop being a baby. You get no bonus points for sticking around.

So writing about babies is a slippery experience: at some point your subjects just disappear. I began this book with a baby in my house. I ended it with a child who shows almost no sign of ever having been a baby. This is the fate of all babies: for millennia now, they have been vanishing.

You might think that these missing babies are a lot like the ones around us. After all, for many thousands of years, babies have been pretty predictable: they come out greasy and looking like Winston Churchill. (Even *before* Winston Churchill, they managed to look like Winston Churchill.) You might think that their stories—their experience of infancy—would be pretty predictable, too: they poop, they burp, they stagger around like early hominids. Who cares where all the babies have gone? Aren't their stories mostly the same? Don't they all smile and spit up? Don't they all refuse to nap?

Their stories are not mostly the same. They are radically, outrageously, wondrously not the same.

This book tells the story of how some people used to think babies work. It also tells the story of how a lot of different people— some developmental psychologists, some physicians, some hunter-gatherers, some you and me—think they work now. I can inform you that among all these stories there is very, very little overlap. Only the species stays the same.

Infancy turns out to be far more varied and far less predictable than we often assume. It's all deviations from the norm—even the developmental milestones themselves aren't normal. Books about babies usually try to iron out all these wrinkles, to make infancy seem smooth, coherent, known. They leave out the dried cow's teats and the skull molding.

But babies are gloriously unkempt. This book, like a newborn, is all wrinkles.

You will have noticed that the subtitle of this book begins with *Suck*. I should really explain.

A book about babies is supposed to begin with *how*. Most books in the parenting section are instruction manuals: they tell you how to get your child to sleep and how to get her into the accredited institution of your choice.

This book doesn't begin with *how* for a very simple reason: it doesn't tell you *how* to do anything. It solves no problems that have to be solved this instant: if your child is screaming *right now*, I cannot help you. Try the next book over.

What this book does do is step back from the problems, so they appear in perspective—perspective being that rarest commodity in parenting. Taking care of an infant is myopic madness: it is hard to make out much past the crib bumper. Your adult life shrinks until it resembles infancy itself: there is nothing remaining but essential bodily processes. I remember whole days from early parenthood when I thought about nothing but green poop. You'd never know there was more to think about infancy. Only by stepping back can you glimpse its strangeness and splendor.

Our approach to infancy is highly parochial: we raise and interpret our children in the hothouse of our present neuroses. We hyperventilate about a handful of things: co-sleeping versus letting them cry it out, slings versus strollers, even spoon-feeding versus finger-feeding. This book tries to break the hothouse glass. It looks forward to how scientists—doctors, developmental psychologists, neuroscientists—are shaping our understanding of how babies work. Theirs is the peer-reviewed baby of tomorrow, and we expert-dependent Western parents will automatically adopt their baby as the new standard model. That's the trade we parents first made almost a century ago: folk knowledge for scientific knowledge.

But not everyone made that trade. In much of the world, babies are understood in ways that seem surpassingly strange to us. This says more about us than them: from our scientific perspective, any model of infancy that isn't at least pseudoscientific looks strange. It's easy to assume that the history of infancy is a story of convergence, with everyone eventually arriving at the same conclusions. It isn't. Like evolution on isolated islands, child-rearing practices are adaptations to specific conditions, which is why people can look at the same greasy newborn and come to so many different conclusions. John Locke insisted on leaky shoes and ice baths. The Beng of West Africa insist on enemas. This book is a mirror held up to a mirror: it looks at all these ways of looking.

It also glances backward to the babies of the past: how people in radically different places and eras—the experts and the parents—once thought babies worked. Not just thought, were *convinced*. And people have believed—and still believe—that babies work in ways even weirder and more outlandish than you might expect: they once thought that colostrum was toxic and that touching a baby could be gravely dangerous. The peer-reviewed baby of tomorrow is just a version among the many that have proliferated over the years and across the globe. And for each society, its version of the baby was *the* baby—the only baby imaginable.

Once you see how many different babies there are and have been, and how many different ways people have insisted were the *only* ways of treating them, all child-rearing advice—a book of *hows*—begins to seem a little suspect. The grown-ups in the room have rarely agreed on how babies work, or even on what they need. Most of the *hows* about babies turn out to have actually been about the adults. You can't write a brief history of babies without also writing a lengthy history of grown-ups.

So this isn't a book about how to do things with babies. It's a book about how babies do things. The subtitle begins with *Suck* for a somewhat improbable reason: this book is actually about sucking. Every new skill in infancy is all-consuming: a newborn will do nothing but suck; a newly smiling baby is as gregarious as a puppy; a baby who can grab something will grab everything; a baby who has just learned how to move will be a blur. But outside of developmental posters and nursing guides, no one says much about these activities: to everyone but the baby, the most visible parts of infancy are the most overlooked.

They shouldn't be. The activities herein—sucking, smiling, touching, toddling—cover a lot of what transpires during infancy: how a baby feeds and consoles herself; how she develops emotionally and socially; how she begins participating in her world; and how she learns to explore it. These are the big puzzle pieces you need to assemble an infant. (It's an incomplete puzzle, of course: the baby in these pages does not cry or sleep. There's so much happening in infancy—the amount of development is so dense—that if you covered everything you'd never get to anything.)

This book is also the backstory of a very specific baby, the raison d'être of the whole thing. He started it. I should really introduce him.

His story begins the instant I broke my elbow biking, a little less than a decade ago. Each of us arrives in this world to a history that precedes us. His was written on the unyielding asphalt in front of a White Castle franchise.

I had to wear a sling on my right arm for the next few months, and at a Fourth of July party, I had to scrape the burnt bits off my hot dog bun with my non-burnt bit-scraping arm and someone standing nearby was audibly impressed at my wrong-armed coordination. It is the only recorded episode of coordination in my life but never mind. We spent the rest of the night on the rooftop talking. She was dating someone else but never mind. It was Independence Day. There were fireworks. It was Independence Day in Chicago, so the fireworks were interrupted by occasional celebratory gunshots, but still: fireworks.

Isaiah was born some four years later. The elbow's doing fine.

Some promises, some disclaimers, some weaseling:

I have tried to avoid telling you what to do. There are already lots of books out there that tell parents how to be parents—I think you can buy them by the cord, like firewood. The only advice in this book is discarded advice: a collection of innumerable core convictions about babies that have turned out to be misguided, comic, harmful, singularly strange, and fitfully useful, sometimes all at once. In place of advice, I offer skepticism. I am not professionally capable of offering anything but skepticism: I am not an expert. I have no training in anything child-hyphenated. I have only a child.

I will not be hawking any child-rearing philosophies. This book doesn't retail yet another parenting approach. It sketches out the bigger story of what's behind a baby's behavior and a parent's anxiety. It illustrates the many different ways babies have been and continue to be raised; it makes clear that there's no *right way* of doing things. In theory, and this is the late-night infomercial part of the introduction, this bigger story should soothe some parental anxieties. It clears some breathing room. It pulls back from the hectic this-then-this nature of infancy and lets parents see babies, and themselves, in full. (Individual results may vary.)

Is that a child-rearing philosophy after all? Maybe.

I have not been fair to everyone and everything. It's not you, it's me: I have my own biases, neuroses, compulsions. None of us are neutral parties here. Some research I found more compelling, some less; some I liked more, some less. (There are probably deep-seated reasons for these preferences, none of which will be explored in this book.)

I have been free and easy about the first-person plural. In some places, I may have gone too far. If necessary, wherever "we" appears, please substitute "You, asshole." (Also, whenever "mothers" appears, it is because the studies are *of* mothers—no one invited the fathers.)

I have tried not to mock the past. This has been hard, because the past sometimes seems like it is begging for it. ("Tickling is bad for children. Sometimes it does serious harm, and it never does any good," the educator Angelo Patri wrote in 1922.) There's no easier target than how people used to take care of children: the parents of the past are beyond empathy. But I will soon be a parent of the past, and I will look preposterous, too. We all will. It's good to remember.

We tell ourselves stories in order to live, Joan Didion wrote. We tell ourselves stories in order to raise our damn children, too. Taking care of a child requires coping with way too much information. To sort through this information, most of us use the time-honored junk mail management strategy: we discard most of it. We just toss it. We only keep the bits that fit our narrative. In each section of this book, the parents and experts of various times and places—pre-Revolutionary France, rural Uganda, us right here—sort a baby's behavior into a story they tell themselves: that it is reasonable to ship freshly made newborns to wet nurses in the countryside, or that babies have to be drilled in order to learn how to sit and walk, or that babies will confuse a pacifier with food. Outside of these times and places, such stories are incomprehensible; they don't make any sense. (A pacifier isn't food! Babies sit up on their own!) But if you're on the inside, they are the only stories that can be comprehended. Nothing else makes sense.

This book is a modest attempt to get outside our own time and place—to try on some other ways of making sense of it all. It's an attempt, at the very least, to look beyond the green poop.

Part One

SUCK

But so soon as ever they are born, nay before they are born, they will suck. For we have found by experience, that while they yet stick fast in the Birth, before they can either cry or breath, they will seize upon the finger extended to them, and suck it.

—William Harvey,
*Anatomical Exercitations
Concerning the Generation
of Living Creatures,* 1651

1

Sucking: A Love Story

From the moment my son was born, the one and only thing he asked of the world was that it give him something to suck. Among the very first photos I took of him, there's a shot that foretells the next half year of his life. His pink-gray hand, wrinkled and fuzzy, takes up almost all the foreground. It's expertly splayed with tension, as if he'd practiced jazz hands in the womb. In the background, slightly out of focus, is his mouth, agape, bobbing for boobie.

He's missing. He has horrible aim.

It's Isaiah's first photo starring him and his new best friend. It is also the beginning of an obsession, his desperate need to get something—anything—into his mouth and suck on it like a toothless vampire bat.

He was designed to be a vampire bat: a newborn is a body led around by a mouth. It is his most sensitive organ, his most useful tool for probing the mysteries of the world—the best way to probe any mystery being to suck on it, of course. When you have a hammer, everything looks like a nail; when you are a baby, everything looks like something to suck. Even nothing is something to suck.

Watch a sleeping or sleepy infant closely and you will often see her mouth slowly, silently, suck on nothing but air.

The typical baby in the parenting books we were reading seemed to be highly discriminating about what she sucked. The parents have tried every bottle; they've bought every brand of pacifier. But the baby accepts no substitutes for the breast. The baby will not be fooled. There's an undercurrent of pride to these stories. Our child is too clever for these tricks, they seem to say. It is difficult having such a smart, loving, prodigiously attached child. But we manage.

This is not what Isaiah was like. He sucked—poorly—on his thumbs and fingers, and—expertly—on dirty laundry, stuffed sheep, our necks, other people's noses. He not only accepted substitutes to the breast; he preferred substitutes. If we had put lumber in his bassinet, he would have sucked it down to driftwood.

Over the years, a lot of people have not dealt well with finding a vampire bat in the nursery.

For more than a century, the sight of a baby sucking—how we have understood it, rationalized it, recoiled from it—has had a weird capacity to inspire and magnify parental anxieties. This is true for sucking in all its manifestations: breast-feeding and wet nursing, thumb-sucking and pacifiers, each of which has its own peculiar and revealing history. Each has left a mark: our continued confusion and ambivalence about pacifiers, and even thumb-sucking, is the legacy of these forgotten battles.

They were fierce, moral battles. Even before Sigmund Freud, a spurious association between sucking and sexuality had planted itself in the minds of parents and doctors. A treatise on childhood diseases from the turn of the century is matter-of-fact about it: "Infants who persist in the habit of sucking always become masturbators." What were these infants sucking? Their thumbs. Such habits in infancy were said to be the seeds of adult delinquency. Mothers who ignored them were woefully naïve: "Even when warned and fully understanding the dangers, they in mistaken kindness, for temporary present good, neglect to provide against certain future evil."

A hundred years later, too much sucking is still distressing. Even Google is aware of our unease: if you type "pacifiers" into the search engine, the first suggestion offered is "pacifiers good or bad."

Babies, of course, have always picked good. "The immediate pleasure value of his mouth activity to the infant himself is easy to see from his reaction," the psychiatrist Margaret Ribble wrote a half century ago in *The Rights of Infants,* the rare book at the time to acknowledge how vital sucking was for babies. "Pleasure is the principle on which he accepts or rejects; at this stage it is his criterion of good and bad, and no Emily Post is going to make him pretend anything different." Those who have looked closely have always been amazed by the deep pleasure and meaning that babies find in sucking. Even Charles Darwin took note of it. "It may be presumed," he wrote with dry affection of his newborn son, William Erasmus Darwin, "that infants feel pleasure whilst sucking and the expression of their swimming eyes seems to show that this is the case." Freud, famously, compared the rapture of sucking—those swimming eyes—to orgasm. I thought Isaiah looked more like he'd slipped under the spell of a powerful opiate, as if he were taking in breast milk derived from a strict diet of poppy-seed cake.

Sucking is the act that governs early infancy. For generations, this fact has been oddly terrifying. It shouldn't be. We know now that sucking is a benevolent dictator: it means well. If anything, it is supremely ironic that sucking has been the subject of so much grief. It's how newborn babies survive. It's almost the only thing they can do. You'd think we'd be happy they were so good at it.

We *Sugentia*

When other mammals arrive in the world, they take advantage of the fact. Shortly after being born, a calf can stand. A deer can run.

A baby can suck.

There are no mammals who arrive in the world to lower expectations than the human newborn. After marsupials, which are suckled in a pouch until they can cross the street, we are the least prepared, least advanced mammals at birth. The amount of time we need before we are truly ready for post-utero life is much closer to eighteen months than to nine. It takes human infants nine months *outside* the womb before they are as capable as a chimpanzee in its first minute of life.

This ineptitude is evolution's blessing and its curse, and to understand why such an arrangement would ever make sense, we have to rewind the tape a couple of million years. The brains of our newly upright ancestors were exploding in size, but the birth canal was shrinking: the new bipedal lifestyle required a narrower pelvis. It was, essentially, "a conflict between walking and thinking."

Bigger heads, smaller openings: no one needed a traffic report to see the bottleneck. "The professions of obstetrics and midwifery probably would not exist," as the anthropologist Wenda Trevathan observes, "were it not for the inherent conflict between the bipedal pelvis and the large-brained infant." There was an imperfect compromise: mothers began to give birth sooner, when the fetus was relatively small-brained. It was the only way to reconcile a big brain with a small pelvis: the bulk of the brain growth—the bulk of what was supposed to happen during gestation—now took place outside the womb, where there were no physical barriers. It was a highly atypical arrangement: the brains of most mammals grow quickly only before birth. The brains of humans quadruple in size *after* birth.

All this meant that gestation, counting the time spent inside the womb and the time outside it, now took approximately forever. There are other mammals that gestate at our achingly slow pace. And there are other mammals that are born helpless. But *Homo sapiens* is the only mammal that's both: we take forever to prepare ourselves and we're still not ready in time.

This paradox is responsible for the extraordinary achievements of the species later in life: among our many questionable singular virtues, it is why we are the only mammal to go to graduate school. But to any mother exhausted after hours of labor, what may be more apparent is how little evolution has willed her at that moment: a mammal—possibly a baby, possibly a naked mole rat—who will not smile at her for well over a month but who will scream from the instant of birth.

He can do almost nothing else. Except suck.

When Carl Linnaeus, the great Swedish naturalist, sorted the natural world into taxonomical categories for his magnum opus *Systema Naturae*—the foundation for how we think and talk about all plants and animals—he threw out the *Quadrupedia* class, which ordered warm-blooded animals by their number of feet, in favor of an order determined by *mammae,* or breasts. Hence, *Mammalia,* or mammals. Early taxonomical classification was an art, not a science, and not always physiologically rigorous: a previous influential scheme had divided animals between the biblical categories of clean and unclean. *Mammalia* was a more scientific classification, but it was still somewhat arbitrary. There are at least a half-dozen characteristics unique to all mammals, including hair. But not all mammals have *mammae:* male horses, for example, as Linnaeus himself admitted, have no breasts.

Linnaeus, as the historian Londa Schiebinger has argued, might have been wiser to move away from form and toward function: rather than focusing—scandalously—on the physiological fact of breasts, he should have directed the eyes of naturalists to what those *mammae* are designed to *do*. This distinction makes so much sense it already exists. In German, mammals aren't named for what they have; they're named for what they do. They are *Säugetiere*—suckling animals. If Linnaeus had made this tweak, we would think of ourselves today not as *Mammalia* but as *Sugentia:* the suckling ones. Sucking is our true mammalian obligation. More than any anatomical fact, it is what connects us to our fellow mammals.

Proof of the nomenclature is in the womb: as *Sugentia,* we give ourselves prenatal hickies. Like children covered with cookie crumbs, newborn babies inadvertently disclose what they have been up to out of sight. They emerge from the womb with their bodies tattooed by compulsive sucking. You can see fetuses contentedly sucking on their extremities on ultrasounds. Infants born without sucking marks on their hands or arms may have already found their thumbs and fingers, which is why a newborn, a creature incapable of holding her head up, is able to find her hands shortly after birth. It's not luck. It's practice. Fetuses are able to move their hands up to their face by the end of the first trimester; by sixteen weeks, they begin sucking their thumbs.

In-utero sucking prepares the fetus for the critical task of nursing. So does swallowing. By the end of gestation, the fetus swallows as much as a liter of amniotic fluid every day just to keep the composition of the uterine environment from falling out of whack. It's proportionally more liquid, compared to body mass, than any of us will ever drink again.

Since taste buds emerge as early as the first trimester, we all once knew what the womb *tastes* like. Fetuses are highly sensitive to taste: experiments conducted decades ago, before the advent of ethical niceties, found that a fetus will swallow more amniotic fluid after the injection of saccharin and less after the introduction of a "noxious" substance. Amniotic fluid has many of the same flavors as breast milk, and newborns may find a measure of comfort in nursing because it makes their new alien world seem a little more like home. They uniformly prefer any odors or flavors they experienced in the womb; they turn toward their mother's amniotic fluid and away from that of another mother. The connection between amniotic fluid and breast milk is the rare bit of continuity from the prenatal to the postnatal world—think of breast milk as the very first transitional object.

As adults, we look at newborns, these befuddled arrivals in our ex-utero world, and we naturally see versions of us—really lousy versions of us. But it may be more fair to see newborns as really good versions of *Sugentia*. They are designed to do what they have to do: their facial muscles, astonishingly well developed, flush milk out of their mothers; their guts thrive on the milk; their teeth arrive tardily, for their mother's sake.

They are born ready to suck. You might think that after the trauma of birth—the physical stress, the astonishment of arriving in a new world—a baby might need some time to get himself together. But the readiness of a newborn to suck, measured by sucking pressure, is at its maximum just after birth. Within six to eight hours afterward, it has already declined.

A newborn is so ready, in fact, that if placed on his mother's chest after birth he will find his way to her breasts on his own. It will take him almost an hour, he will have been guided almost exclusively by his nose, but he will make it. It's an extraordinary act of blind will. Newborns are comically nearsighted—approximately 20/400—but their sense of smell is fully developed and they suss out their surroundings with it. At a week, breast-fed babies identify their mother by smell: in experiments, they consistently turn their heads toward her breast pad and away from the pad of another lactating mother. (A week later, they identify their mother's *armpit* odor.) When newborns are placed on their mother's chest between a washed and an unwashed breast, they consistently choose the unwashed breast—the side with her scent.

On the spectrum of postpartum nursing competence, we are closer to rats, who can't nurse if they can't smell, than farm animals like sheep and cattle, who can wobble their way, eyes open, to the nipple. Precisely what we do when we get there had long been a mystery. Only twenty-five years ago, a paper on the anatomy of sucking began by saying, "It is clear from recent lay and professional texts that there is much confusion as to the precise nature of infant sucking."

There's a good reason for this: it is genuinely confusing. What babies do is not what the word "sucking" leads you to expect: babies pump out milk not so much by sucking as by *squeezing*. The infant squeezes the breast between his tongue and palate, elongating the nipple and compressing it with wavelike motions of his tongue; this expresses the milk, which is then swept away for swallowing. Suction doesn't actually propel the milk from the breast; it just keeps the nipple in place. (Our idea of how breast-feeding works—an infant *sucking* milk out of the breast—is actually how bottle-feeding works.)

Scientists have now studied breast-feeding in millisecond-level detail. With mathematical rigor, researchers have tried to calculate norms for swallow rates, suck runs, efficiencies of feeding, rhythmic stability—every aspect of a suck or swallow. Having hyperdetailed data enables earlier diagnoses of neurological or feeding disorders in high-risk infants. And feeding disorders are common: feeding is complicated. In order to nurse successfully, a newborn has to coordinate sucking and swallowing with breathing: a few milliseconds off in any direction and the whole apparatus malfunctions. If a baby breathes in milk, he chokes; if he swallows air, he gets gas. Eating is a remarkable achievement, the most complex motor act a newborn can accomplish. In approximately a tenth of a second, a litany of crucial openings and flaps are covered, raised, lowered, squeezed—all to prevent liquid from dripping into the lungs. Things can go wrong at any point in this process, for almost any reason. It is a physiological marvel that all infants don't have pneumonia from sodden lungs.

This is what's happening behind a baby's chugging cheeks. We notice almost none of it. We *can't* notice it. Sucking is a high-wire balancing act performed only behind a curtain. It's a marvel that fails to seem marvelous.

2

Lactation and Its Discontents

If sucking fails to seem marvelous, what does seem marvelous is what it yields: mother's milk.

We know it is marvelous because we are constantly told that it is marvelous. It seems as if every month brings another study showing that breast milk is what Ponce de León should have been searching for. Breast-feeding has been shown to improve the health of mother *and* child—the mother is less likely to have high blood pressure or cardiovascular disease later in life; the child is less likely to be obese, to have asthma, to have celiac disease. (Of course, these studies show correlation, not causation; they don't prove that the breast milk is what did it.)

In the most remarkable recent study—remarkable because it showed how wondrously byzantine the workings of breast milk are—scientists looked at a bacterium that lines the intestines of breast-fed babies. It's long been known that the gut flora of breast-fed infants is different from the flora of infants fed formula, and that, perhaps because of this, breast milk protects against diarrheal diseases. It's also known that a mysteriously large fraction of breast milk—some 21 percent of it, made up of complex sugars—is indigestible by infants. This study put these pieces together and concluded that the indigestible fraction of breast milk isn't meant to feed the infant but to feed this particular bacterium in the infant's gut. In other words, the indigestible component of breast milk is there to nurture the protective *bacterium*. It is as if, one of the scientists said, "mothers are recruiting another life-form to baby-sit their baby."

That is where we are now: the scientists no longer bother to prove that breast milk is good for you. They now prove that breast milk is good for the creatures that live inside you. The public service advertisement writes itself: *Nourish Your Baby's Gut Bacterium: Breast-Feed*.

These are fine-grade studies. In developing countries, you don't need fine-grade studies to make a compelling case for breast-feeding. You don't need to track health outcomes for decades. You can do it in a few months: whether a newborn is nursed is the key variable between survival and early death. A recent report in *The Lancet* concluded that 1.4 million deaths worldwide could be prevented by breast-feeding. In countries with poor sanitation and hot climates—toxic water, rapid spoilage—it is hard to overstate the importance of breast-feeding to reducing the rate of infant mortality.

The next question is obvious: what about everywhere else? In developed countries, where the immediate health risks of not breast-feeding are extremely low, how much does breast-feeding matter? All of a sudden, a lot of people are asking that question. The cover of *The Atlantic*: "The Case Against Breast-feeding." *Harper's*: "The Tyranny of Breast-feeding." These skeptics aren't arguing that breast isn't better, at least not for the most part. They're arguing that breast might only be a *tiny* bit better, that its benefits are being massively oversold, that overworked women shouldn't be told that formula is a menace to their children (and their motherhood).

This argument won't be settled. It *can't* be settled. It isn't really a debate about breast milk versus formula. It's a debate about the obligations and responsibilities of mothers in a working-mother world; about when science is dispositive and when it is shaky; and about whether women will always be biology first and everything else second.

It's also, in our world of billable hours, about how much time breast-feeding demands. Blame the hominid ancestors: because early humans were constantly in contact with their infants, milk was always available—there was no need for it to be a complete meal, or even especially filling. So humans, like other primates, cut the fat with water: breast milk is 3 to 5 percent fat, around 90 percent water. If we nursed like hooded seals, we wouldn't be having this conversation: the hooded seal, which lactates for less than a week, has milk with a fat content of 60 percent. (Some perspective: premium ice cream comes in around 15 percent.) Baby hooded seals can gain fifty pounds in several days. The tree shrew has extremely high-fat milk to compensate for the few times it returns to the nest to feed its young. The hippopotamus nurses once a week. Neither hooded seals nor tree shrews nor hippopotamuses use breast pumps.

But let's stipulate, for the sake of argument, that breast milk is in fact the pixie dust of life. This raises an awkward question: since newborns need to suck in order to survive, and since they spend much of their prenatal time practicing sucking and swallowing, and since they have an all-consuming drive to suck, and since breast milk itself is something of a magical elixir—when these babies are finally at the *mammae,* why do they screw it up so much?

The (Sometimes Not So Much) Wonder of Breast-Feeding

The problem was not a lack of enthusiasm. Calling it enthusiasm sells Isaiah short, really. It was closer to fanaticism. There are few analogies outside babyhood for this intensity of focus—later in life, we are easily distracted by hot chocolate, or Angry Birds, or updating our relationship status. The world-obliterating significance of sucking is so alien to adult life that when psychologists began contemplating its power a century ago, they were unable to conceive of it as normal or healthy. Instead, they immediately drew a parallel to drug addiction; many concluded that sublimated sucking lay behind all addiction. In 1925, the American psychologist James Mursell went so far as to argue that "the drive behind the smoking habit cannot be due to the specific effects of tobacco as a drug, for these are negligible in any case." Any effects of alcohol and tobacco, he concluded, are "largely fictitious." Sucking was the true menace.

I could handle the sucking on necks and noses. The stuffed sheep could handle themselves. Anything more nutritive was Anya's responsibility. She was the girl on the rooftop and this was a mess of her own sad making: she'd married me.

Unlike Mursell, she wasn't worried about what Isaiah's sucking drive meant for him. She was worried about what it meant for her. Usually, it meant pain: sometimes what people kept calling "good" pain (from the shock of Isaiah latching on); sometimes faux good (which seemed "good" until it suddenly did not); sometimes just bad (which was well-named).

So when she felt a new small pain, about three weeks in, it wasn't cause for alarm. It was cause *not* to be alarmed, really. It was just more of the same. The natural state of new parenthood—an addled, heavy-lidded blind hope—makes it tough to see beyond the next diaper. It's hard to worry about what's ahead when you don't even know where you are.

A few days after the new pain appeared, Isaiah stopped nursing on that side. The doctor diagnosed mastitis and put Anya on the sort of schedule that makes normal postpartum sleeplessness seem like a spa retreat: she fed (an hour), pumped (an hour), ate or slept or tried to remember who she was (an hour), after which

Isaiah was hungry again, so she fed, pumped, ripped the mobile from the ceiling. It didn't help much. Despite all these warning signs, it was not until New Year's Eve—when the emergency room doctor, the first doctor to recognize what was actually wrong, said "oh" and "yep" and "next room" and "scalpel"—that I realized this pain wasn't just more of the same.

Later, after the abscess was drained, Anya said, "You don't even know when something is seriously wrong because it all feels so wrong to begin with."

Somewhere along the line, the fact that breast-feeding is natural—the rare thing that both sides of the nursing wars agree on—bled into the belief that it is instinctive. How we think about the act has not helped. Breast-feeding is cast as a series of iconic, Madonna-esque images. It's portrayed as the embodiment of maternal love, a mother-and-child idyll, a sanctified experience. The language for it is not much more down-to-earth. The psychiatrist Daniel Stern's well-meaning description of a baby finding her way to the nipple is typical: "A well-choreographed pas de deux." From such portrayals, it is hard not to conclude that nursing is just doing what comes naturally, a physical expression of a mental state. And if breast-feeding is simply an expression of maternal love, who can fail? If you were to fail, what would that mean? "The only emotional experiences mentioned in the 250-page American Academy of Pediatrics breast-feeding guide are 'joy,' 'fulfillment,' and 'emotional communication,'" as the critic Rebecca Kukla has observed.

There are dissenting voices out there. Each year brings more books to be shelved in the new genre of deeply unsentimental motherhood. But partly because of cultural whitewashing, partly because breast-feeding is our mammalian inheritance, it is too easy to assume that it ought to be intuitive. When Anya's mother saw a copy of *The Nursing Mother's Companion* in her hospital room, she said, indignantly, "What is there to write a book about? You just put the child to your breast!" *The Nursing Mother's Companion* was a scam, she thought, and I suspect there's a small part of us all that thinks she's right, or that at least thinks she ought to be right. How can such an important act *not* be intuitive? You just put the child to your breast! Gorillas don't need *The Nursing Mother's Companion*.

Here's what happened when a gorilla raised in a zoo nursed her newborn for the first time: she cradled him, nestling his head against her chest, and carefully guided her nipple into the back of his head. The first chimpanzee to give birth after being raised in captivity had both her infants die of starvation. They never figured out how to nurse; she never figured out how to teach them. Another gorilla raised in captivity learned how to nurse by watching videos of other gorillas nursing. A lot of primates, it turns out, could use a copy of *The Nursing Mother's Companion*.

In most mammals, the process of nursing *is* largely instinctive and reflexive, as Derrick and Eleanore Jelliffe note in their wonderfully titled book, *Human Milk in the Modern World*. But humans and other primates are not among those lucky mammals. For us, breast-feeding "is partly based on learned behaviour, on information obtained from experienced females, mainly by observation and example." Primates raised in zoos are sometimes unable to obtain this information; they are sometimes helpless without it. A gorilla living in an unnatural environment can't do what comes naturally. Neither can we. Walled off from our neighbors, far from our relatives, cut off from previous generations, we sometimes can't find the other gorillas either.

By the end, Anya had collected a full set of lactation consultants: at the hospital, Isaiah's pediatrician, her doctor's. They all seemed to be hounded by the fear that at any second, and possibly this very second, she might switch to formula and forget all the pain and confusion, and that looming, ever-present threat spurred them into hyperactive helpfulness, a torrent of run-on, semi-solicited advice, as if they believed that the moment they stopped talking Anya would immediately log on to the Enfamil site for coupons. They were also invaluable. They were her gorillas.

The abscess was relatively small and it healed fully. Several months afterward, Anya began feeding Isaiah on the affected side, although there was never much milk there again and he'd often suck for only a few seconds before pulling away with a wounded cry. She somehow nursed him for six months anyway, the official WHO-recommended goal for exclusive breast-feeding.

Isaiah was like a dog who'd learned to run on only three legs. It wasn't the usual way of fetching sticks. But he got to them. And after we heard enough stories from friends and suddenly intimate strangers, we realized what should have been obvious from the beginning: every baby's a three-legged dog. It is the rare act of breast-feeding that is not screwed up in its own special way.

The care of every breast-feeding woman is screwed up in its own special way, too.

Long ago, the study of lactation slipped between the disciplinary cracks of the medical profession. There are no doctors ultimately responsible for a nursing woman's breasts. In fact, until doctors began to outsource breast-feeding to lactation consultants, no one was responsible for it at all. It's an ironic turn of events. Back when pediatrics was still a young, scrappy medical specialty—a little over a century ago—it booted midwives from their historic responsibility of caring for infants and infant feeding, arguing that new mothers and their babies required the attention of trained medical professionals. In the early 1900s, pediatricians staked their authority on their knowledge of infant feeding: many counseled mothers to use byzantine mathematical equations for mixing formula, a "percentage" system so complicated that mothers, who had rarely consulted physicians about feeding before, were now unable to feed their babies without asking for help.

Decades later, when nursing regained popularity, physicians inverted the logic they'd used against midwives: they decided that lactating women were best treated by people who *weren't* doctors.

This is why many nursing women, if they end up in the hospital, end up feeling like a hot potato. *You take her. No, you take her.* For the last century, the care of breast-feeding women has been nothing if not a very long game of hot potato.

The day after Anya's abscess was drained, she was sent to the emergency room by her doctor, who was worried the incision was infected. We spent New Year's Day there and after an afternoon's worth of opinions and grimaces, a young resident, who was also a mother, told us that she thought the incision was fine after all. The pus that doctors thought they were seeing wasn't pus, she said: it was milk. She then said, explicitly, that we shouldn't be there at all; that none of the doctors in the hospital really knew anything about lactating women, and if Anya wanted to see someone who actually knew something, she should see a doctor in another town altogether, someone who was a pediatrician and a lactation consultant, a rare combination. It was a revealing moment, not just because doctors rarely admit to not being the last word on a subject, but because of where we were: Yale–New Haven Hospital, a hospital routinely ranked among the very best, a place where doctors are presumed to know everything they could know. It's an institution patients are transferred to, not from.

We didn't take the resident's advice. But she wasn't wrong. Shortly thereafter, Anya had a follow-up appointment with a senior breast surgeon. He told her, with the no-nonsense authority of his white lab coat, that if she only nursed on one side, all of her milk would dry up. He was wrong and it wasn't a minor mistake: it was flagrant; it should have been embarrassing. But he didn't know, and he didn't care, and he didn't need to: his practice was built on breast cancer, not breast-feeding. For him, as for many doctors, lactation was an anomaly, a temporary complication—it, too, would pass.

So it shouldn't come as a shock that some of the most significant medical work on the anatomy of the lactating breast was
done over 150 years ago, when the British surgeon Astley Cooper
dissected the breasts of women who'd died while lactating. Pla-
cental mammals like humans arrive in the world in risky, high-
stakes fashion: we emerge from the womb in urgent need of
succor. Nursing seems like the sort of bedrock process that would
have been well mapped out by now. But only in recent years have
doctors and lactation consultants sketched out an account that
corresponds to the actual activity. Not long ago, descriptions of
the lactating woman and the sucking infant were closer to medieval cartography: they were more figurative than literal.

The Persnickety Newborn

We don't just misunderstand the mother's role in the "pas de
deux" of nursing. We misunderstand the baby's role, too.

Breast-feeding is usually described as a purely reflexive process: first, the rooting reflex, which directs the head toward the
nipple; then the lip reflex, which guides the mouth; and finally,
the sucking and swallowing reflexes, which finish the job. The
newborn himself is a bystander, who watches from the couch
while his reflexes do all the work. It's instinct in action.

But the poignant lesson from apes struggling to nurse is that
many things that would seem instinctive aren't. And recent research on newborns concludes that a lot of things we thought
were reflexes aren't. Newborn behavior cannot be reduced to
reflexes, according to Claes von Hofsten, a professor at Uppsala
University in Sweden who has done seminal work on how babies
interact with their environment. "This is not to deny that neonates have reflexes. They have them just like adults. However, to
describe normal neonatal behavior as expressions of reflexes is
wrong." The current consensus, in fact, is that rather than a

bundle of reflexes, newborns are highly motivated, goal-oriented beings.

To anyone outside the discipline of developmental psychology—to parents—such a consensus seems like a stretch. Everyone knows a newborn is actually a flailing, flopping, cross-eyed, still-fuzzy mess. And for a very long time, this was the consensus of psychologists, too: they assumed that newborns were, when you got down to it, pretty much just potted plants. And for generations of parents, this idea shaped how they saw their new children: passive, inert, in need of occasional refreshment.

This assumption—the potted-plant paradigm—originates in the work of Freud and Jean Piaget, who painted the backdrop for how we think about babies today but whose work was radically different. In fin de siècle Vienna, Freud established infancy as a separate stage of life, deriving his theories from adults; back then, almost no one thought infants would have anything to say about infancy. Decades later in Geneva, Piaget sparked the modern science of infancy by actually studying infants. Piaget's research methods were, by and large, ingeniously ordinary: for hours each day, years on end, he observed infants and children, including his own, accumulating the minute material from which he'd derive *his* theories. Piaget's work proved that babies were not black boxes—with close observation, an abundance of patience, and some clever do-it-yourself experiments, scientists could find a way to see into their lives. Infants actually had a lot to say about infancy.

But neither Freud nor Piaget thought a just-born baby was aware enough to interact with the world around him. Freud believed that newborns were oblivious to anything outside of themselves; he compared them to a baby bird still in the shell, physically separated from the world. Piaget thought that at birth babies aren't separated from the world but melded with it: there was no space between the baby and his environment. He couldn't put enough distance between himself and the world to perceive it: when you're underwater, you can't tell that you're wet.

Because it was assumed babies couldn't interact with their environment, everything they could do was counted as either a reflex or a meaningless, peripheral response. Piaget interpreted sucking, for example, as an instinctive, essentially random response for at least the first month of life.

This was eminently plausible. It happened to be wrong. Scientists now believe newborns are highly attuned and responsive to the world from birth. Take rooting—the way an infant sweeps his head back and forth in search of the nipple. It classically goes by the term "rooting reflex" and it is a core, almost inviolate feature of the infant. A baby who can't root is a baby who can't eat. It's the last thing to be thrown overboard. Even babies with severe brain damage, even those born without a cortex, root.

But the rooting reflex is misnamed. When psychologists tracked how babies react to touching their own faces versus how they react to someone or something else touching their faces, they discovered a striking discrepancy. When newborns touch their own faces, they don't bother looking around for a nipple. But when someone else touches their face, they root frantically. A true reflex wouldn't discriminate between these situations: both would produce the same response. The fact that newborns have different responses suggests that they have a primitive sense of self: they can distinguish between what they do and what someone else does.

Like rooting, sucking was long assumed to be a reflex, as Piaget contended. But there is now good evidence that sucking is not only not a reflex but, as von Hofsten writes, "a very complex behavior with very little in common with reflexes." Because the amount of breast milk available constantly changes—a gush during letdown, a trickle at the end—a nursing infant must constantly modulate how he sucks (and how he swallows and breathes). Recent research on the biomechanics of sucking suggests that infants calibrate and recalibrate their sucking pressure according to the amount of milk currently in their mouth. Occasionally, there's so much milk that they forget to breathe; they're too busy swallowing. Even babies with healthy rhythms of sucking and swallowing will pant when the flow of milk slows. Like a marathoner after a race, they have been waiting to take a full breath.

These early achievements matter: the flexibility and responsiveness required for nursing is strong evidence that newborns are meant to work in coordination with the world. They can't be oblivious to it. Like their mothers, they can't coast on their instincts.

Mucus, Semen, Colostrum

But there are some things that newborns just *know.* They know that, when they finally arrive at the breast after birth, they are not supposed to suck. They are supposed to lick. If left undisturbed, they will lick for as long as five minutes before they begin trying to suck. If given a bottle instead of a breast, they are unfazed: they lick the nipple of the bottle.

The compulsive licking triggers a life-saving domino effect. But it isn't the baby who's saved. It's the mother. In response to all this licking, her body releases a flood of the hormone oxytocin, which contracts the uterus and propels the placenta out. The massive surge of oxytocin helps stanch postpartum bleeding; it may save lives in the process. The lavish licking by a newborn is his first Mother's Day gift.

But he isn't selfless. Early licking and sucking stimulates prolactin, the hormone that ramps up milk production. Prolactin is what makes lactation into a demand-driven system, or as the biological anthropologist Sarah Hrdy puts it, "a perpetual buyer's market." As long as someone is buying, as long as a baby is sucking, there's plenty of prolactin; as soon as a baby stops, its levels plummet. "The parenting hormone," as prolactin is often called, has fascinated scientists since it was discovered in the 1930s. Men living with pregnant women have significantly higher prolactin levels. Most exotically, the hormone produces a condition in wolves and dogs called pseudopregnancy: after sudden spikes in prolactin, animals who aren't and weren't pregnant begin lactating. In Hrdy's book *Mother Nature,* there's a photograph of a pseudopregnant Jack Russell terrier nursing a litter of kittens: driven species-blind by her surging maternal instinct, she had chased off their mother and adopted them.

The stimulation of prolactin is not an efficient process, but newborns at the breast are remarkably patient. Like someone turning the crank of a Model T, they understand they're not going anywhere right away. And for the first few days what they get for their trouble is very nearly nothing at all: a few sticky, sickly-looking golden droplets—colostrum.

Today colostrum is understood as precious and vital, a brilliant bit of vamping that the body does while waiting for the milk to arrive. Rich with proteins and minerals, colostrum is critical life-support in the vulnerable first few days after birth. Packed with immunoglobulins, it is effectively a baby's first vaccination. It purges any meconium, the prenatal poop.

This is our modern, medicalized understanding, at least. It is exactly the opposite of how colostrum was often seen in the past and how it is still regarded in many preindustrial societies: a potent, dangerous substance that should never be fed to newborns. For many different cultures and in many different eras, colostrum, the best food imaginable for newborns, was fervently believed to be the worst.

Colostrum suffered from a paradoxical problem: it was so healthy it looked unhealthy. Yellow and viscous, it is less like breast milk than some sort of undesirable effluvia. European doctors forbade it for centuries. Women in much of South Asia considered it pus; Hindu priests proscribed it from being fed to newborns. (They recommended butter and honey instead.) Mothers in many traditional cultures still rely on other foods until their colostrum dissipates. In some cultures, a baby is fed by a lactating relative until the mother's milk comes in.

Colostrum has been found guilty of every crime. It has been withheld because consuming it would make the child stupid. It has been withheld because it was contaminated by semen, which would make the newborn sick—especially if the semen was from someone other than the father. It's been withheld because deprivation would prepare the baby for famine later in life. It's been withheld because it was stale and rancid. How could it still be fresh after being stored in the breasts for the entire pregnancy? No one knew that colostrum ends up in every newborn, no matter how far he's kept from it: shortly after birth, the fluid can be expressed from a baby's microscopic breasts. This neonatal colostrum is a byproduct of the maternal lactation hormones filtering into the fetus, creating a temporary condition of really-early-onset puberty.

Ancient Rome prized colostrum; it was even the subject of an endearment: *Meum mel, meum cor, mea colostra*—my honey, my heart, my colostrum. But Rome was the exception. After Rome, European physicians would not agree on the worth of colostrum until well into the 1700s. Before then, doctors tried to avoid "contaminating" the child with colostrum: they often forbade mothers from nursing until as long as a month after delivery. This delay caused numerous unintended consequences, not least the pressing problem of how to keep up the milk supply—and what to do with all that milk in the meantime. Women were forced to improvise, making do with primitive breast pumps called sucking glasses, or an older child, or even puppies. These methods endangered both mother and baby: clogged and infected breasts were a significant cause of maternal mortality for centuries. Women desperately needed not more but less doctoring. The massive reduction in the infant mortality rate in Britain over the eighteenth century is partly due to physicians letting mothers keep and nurse their babies from birth—in other words, to physicians staying out of the way.

The connecting thread in the many reactions to colostrum is *weirdness*. Colostrum seems to inspire it. It's easy to see why: shortly after it appears, it disappears; it resembles mucus more than milk; it leaks out in such minuscule quantities that it hardly seems like it could matter. Colostrum seems destined to be misinterpreted. Its only saving grace is its brevity, which contains the damage. After colostrum runs out, you'd think that the course of infant feeding would straighten up and become a lot more sensible.

For modern parents, it has. The myriad and bizarre ways in which babies were once fed have shrunk to a couple: breast or bottle. Around the turn of the twentieth century, the question of what to feed newborns became a simple tug-of-war between breast milk and formula, and things haven't changed much since. In Anya's maternity ward, there were colorful banners comparing the hospital's rate of exclusively breast-fed babies with the rates of other cities and countries. When we took Isaiah home, we walked under those banners with bags of new baby paraphernalia we'd just received: formula samples, a formula company diaper bag, coupons for yet more formula. The advice was contradictory but the options were clear.

It's easy to assume the existence of options for infant feeding is new: that as long as babies have been fed, they have been breast-fed by their mothers, and that we live in a unique and unnatural age because we now have a choice. That's why formula can seem like a violation of the natural order of things and the option of bottle-feeding like a blessing and a curse: on the one hand, there's the convenience; on the other hand, there's the whole betrayal of millennia of human history.

This is the story we usually tell ourselves, at least in some variation. But it turns out not to be a true story. In fact, in some significant ways the conversations parents have today about infant feeding are *less* complicated than they have ever been. People have been struggling with the question of what to feed newborns for thousands of years, and many of them arrived at solutions we never considered for Isaiah: having him suckle at the teats of a goat or spooning him full of bowls of puréed bread and water, topped off with some beer.

It doesn't take much time to compile a lengthy, shocking list of ways in which we *don't* try to feed babies anymore: sliding an infant underneath a goat—on a specially constructed wooden tray—is only the beginning. In almost every time and place before ours, mothers considered an option no one considers today: someone else's bosom. You could argue that the most novel fact about modern infant feeding isn't the existence of formula but the disappearance of wet nurses. The profession died out within the last century, but it already has an ancient ring to it. Formula and artificial nipples have been so successful that they have erased wet nursing's uncomfortably close history. But rubber nipples are the real anomaly. For millennia, a bottle came in the shape of a wet nurse.

3

The Very Weird, Perfectly Normal Story of Wet Nursing (Plus, Some Goat Milk and Beer)

There are records of wet nursing in the earliest stirrings of civilization. In ancient Mesopotamia, contracts were drawn up between wet nurses and their employers. In ancient Egypt and Greece and Rome, slaves were commonly conscripted into nursing the infants of their owners.

The modern history of wet nursing begins in the late Middle Ages, when shopkeepers and artisans—working people—started to hire wet nurses. The wealthy had already employed wet nurses for centuries—in fact, the higher birthrates recorded among the aristocracy were probably made possible by wet nursing. The burgeoning middle class—not just artisans and shopkeepers, but lawyers, clergymen, doctors, and scholars, the strata between the aristocracy and the impoverished—democratized the practice. Over the centuries, it became increasingly common. Most infants may have still been breast-fed by their mothers, but by the late eighteenth century in Britain it was a social norm for middle-class families to hire a wet nurse. Wealthy British women so rarely nursed their own children that those who did sometimes had it noted on their tombstone.

And in Britain, wet nursing was actually less popular than in much of the Continent. The fault line was religious. The Catholic Church effectively encouraged wet nursing. Since lactation was a form of birth control, the Church, which taught that the only legitimate purpose of sex was procreation, banned intercourse for as long as a woman was nursing. The edict was, not surprisingly, widely ignored. But there was a way to comply with it: if a woman stopped nursing altogether and hired a wet nurse, it would be impossible for her to violate the ban—or to be suspected of violating the ban. Having mothers *not* nurse was the only way the Church could be sure that mothers were not having sex while nursing. It was a very clever doctrinal end-run, as long as everyone involved avoided thinking too hard about the sex lives of the wet nurses.

The Protestant sects took the opposite tack: they regarded wet nursing as a divine betrayal. Women who did not nurse were said to be selfish and even evil; ministers preached against their wickedness. The Puritans in particular were the most impassioned supporters of maternal breast-feeding who can be imagined. Their evangelical rhetoric was often just an elaborate nursing metaphor: ministers were the breasts of God, and on Sundays parishioners would "suck the breast while it is open." Eternal life was like "being laid in the bosom of Christ, when sucking the breasts of the grace of Christ." Weaning was losing faith. The great Puritan minister Cotton Mather wrote that women who don't breast-feed "are dead while they live." But Mather, despite the sort of pronouncements that make La Leche League look wishy-washy, had at least some of his own children wet-nursed, and colonial America would hardly represent a clean break from the practice. In the eighteenth century, human milk was already said to be "the most frequently advertised commodity" in the land.

Meanwhile, Paris, 1780: of 21,000 babies born, 700 were breast-fed by their own mothers. The other 20,000 were outsourced to other bosoms. A fortunate few were suckled by wet nurses who lived with or nearby the family, but the overwhelming majority of new Parisians—around three out of every four—were sent to the country to be wet-nursed. It was said that Paris was a city without babies. They were all in the countryside, rooting for a nipple.

Far from Paris and the supervision of their families, the babies struggled. Sometimes they struggled just to find a nipple. Destitute wet nurses often took in more babies than they were able to nurse; many infants were weaned. Some fathers sent their children to be nursed and then stopped paying for their care. The arrangements were so clumsy that babies were lost or switched. Some were sent back to Paris without documents, stripped of an identity, effectively orphans: no one knew who their parents were. Wet nursing per se was not the problem: if the nurse lived in the family's home, a wet-nursed infant did as well as a baby fed by his own mother. Babies sent to the countryside, however, had a mortality rate twice as high.

This wholesale exporting of newborns is stubbornly difficult for us to imagine today, and almost impossible to empathize with. The bare facts of it have sparked debates about the existence of the most basic human qualities. An academic reviewing a survey of the French wet nursing business is reduced to plaintive cries. His sentiments are typical: "What of motherly love? What of maternal-infant bonding? What of that whole complex of close sentiments spun about the child that we associate with the nuclear family?"

Questions like these aren't really about the families of eighteenth-century France. They're about us, and our continuing inability to understand a culture whose choices seem at worst merciless or at best cruelly calculating. Until recently, few people bothered trying to understand: almost all the wet nursing literature came down to condemnation, either of the wet nurses (immoral, debauched, greedy) or the mothers (vain, heartless, scheming). The women involved might insist that things were more complicated.

In dense urban areas like Paris, where rents were extremely high and wages low, a woman's income was often indispensable—and the arrival of a baby threatened to upset a family's delicate financial balance. For these women, wet nursing was a reasonable response to a desperate plight: nurse the baby at home and put everyone's future at risk; or have the baby wet-nursed and save at least the rest of the family. The few wet nurses in the cities, who could be inspected and approved, were employed by the wealthy. Middle- and lower-class families were forced to send their children to distant provinces. "Working mothers sent their babies farther and farther away," as Sarah Hrdy writes, "not to get them legally killed but primarily in quest of affordable care." Wet nursing, which can seem so alien that it throws fundamental things like maternal love into question, was actually a response to a very modern problem: a time-consuming newborn, a mother who needs to work, and a really lousy set of day-care options.

Women were sometimes pressured to use wet nurses by their husbands, who would be cut off from conjugal pleasures otherwise, at least according to the Church's edict. And doctors inadvertently pushed babies into the bosoms of wet nurses: by forbidding breast-feeding until after colostrum dissipated, they made it far more likely to fail. Physicians sometimes made new mothers wait for days before they were allowed to see their newborn child. It was safer to keep them separated, they believed, just in case. (Besides colostrum, there was a bevy of other reasons: that a mother's first milk was impure, that she shouldn't feed until she was fully recovered from giving birth, that a baby shouldn't eat if he was still excreting meconium.) The barriers placed between mother and child help explain why so many mothers were willing to farm out their newborns: they were never really theirs.

Not nursing meant that mothers were far more likely to get pregnant, which meant they'd have another newborn to send to a wet nurse, which meant they were far more likely to get pregnant. Family planning at the time resembled nothing so much as an endless round of "My Name Is Yon Yonson." Near-perpetual pregnancy was a crushing, deadly plight. Working women—the same women who often couldn't afford to nurse their children because their labor was too valuable—regularly gave birth, over the course of a lifetime, to a staggering twelve to sixteen children. Wealthier women who never nursed were sometimes pregnant almost every year for the first couple of decades of marriage.

In France or elsewhere, the wet nurse was scarcely better off. If she had milk to sell, she almost certainly had her own child to feed: if she neglected her child, he might die; if she neglected the baby she was wet-nursing, *he* might die, and she'd lose the income he was bringing in. Although well-fed, well-rested women can nurse multiple infants, wet nurses were rarely either. To make matters worse, doctors insisted, erroneously, that women who'd been nursing for over six months produced milk that was "too old" for newborns. The insistence on "new milk" meant that a wet nurse would almost always have a newborn child of her own. In the United States, where wet nurses typically lived with the family and rarely were allowed to bring their own child along, the ghostly presence of the abandoned child hovered over the household. Wet nurses were constantly accused of putting their own children first, of sneaking back to see them or nursing them on the side—in short, of violating the implicit contract of wet nursing: to forget about your own child. Everyone knew that this other child existed: books on how to choose a wet nurse actually advised parents to verify the health of a candidate by examining her newborn. A mother without a healthy infant had no credentials; her milk was automatically suspect. It was rare to hire a wet nurse whose baby had died. Her baby was supposed to die *after* she was hired.

The doctors involved knew the percentages. "By the sacrifice of the infant of the poor woman, the offspring of the wealthy will be preserved," a nineteenth-century physician described the business. Less matter-of-fact is the Irish novelist George Moore's account in his novel *Esther Waters* of a wet nurse forbidden to see her sick child:

What about them two that died? When you spoke first I thought you meant two of your own children, but the housemaid told me that they was the children of the two wet-nurses you had before me, them whose milk didn't suit your baby. It is our babies that die, it is life for a life; more than that, two lives for a life and now the life of my little boy is asked for.

From the employer's perspective, a good wet nurse should be happy to sacrifice her own "little boy" and happy to love other children in his place, especially the children of her employer—who were, of course, the reason her own child was sacrificed. It was an impossible bind, but the ideal wet nurse was always a contradictory combination of characteristics: she should have children and be celibate. She should be calm and unemotional and also empathetic and playful. She should not be poor but she should be poor enough to need work as a wet nurse. She should be, in short, the virgin-whore of the nursery.

At its peak, wet nursing was attacked from within. The philosopher Jean-Jacques Rousseau, a man who had abandoned all five of his children to orphanages, seized upon wet nursing as the genesis of all social problems. In *Emile,* his treatise on education, Rousseau proclaimed that breast-feeding was the first principle of any healthy society: "Let mothers deign to nurse their children, morals will reform themselves, nature's sentiments will be awakened in every heart, the state will be repeopled." The mothers would nurse the new state into creation. After the French Revolution, the legislative assembly cut off state support from all mothers who employed wet nurses. It passed a law decreeing that during national festivals, nursing mothers would be honored immediately after public officials. The new citizens of the Republic, it was said, "must imbibe the principles of the constitution from infancy." A year later, in 1794, Prussia, clearly imbibing the same rhetoric, legally *required* all mothers to breast-feed.

Across Europe, governments slowly took notice of their staggeringly high infant mortality rates; many responded by making maternal breast-feeding a priority of the state. In the late nineteenth century, Germany launched a campaign to move infants back to the bosom of their mothers. In Munich in 1877, only 14 percent of infants were nursed by their mothers; by 1933, 91 percent were. In Britain, after over a third of all recruits for the Boer War were deemed too weak for service, the government began to argue that mothers should breast-feed for the sake of the Empire itself.

Ironically, as maternal breast-feeding became more popular, so did artificial feeding. Food companies and doctors trumpeted highly hypothetical advances in infant food. "No More Wet Nurses" shouted the advertisements for Liebig's in 1869, the first baby food marketed in the United States. But much of this new "scientific" food was either deadly on its own or became deadly when mixed with spoiled cow's milk, an endemic problem. In London in 1900, a third of milk sold to the public was contaminated with pus from diseased udders. Physicians were forced to recommend wet nursing if an infant's health was endangered. Many resented it, though: "A wet nurse is one-quarter cow and three-quarters devil," a Chicago pediatrician told his colleagues.

At the turn of the twentieth century, wet nurses were still hired, especially by upper-class families. But the role of wet nurses was already diminishing, and by the end of World War I, the business would shrink further, across Europe and the United States. It did not vanish—the 1927 *Infant Care,* the U.S. Children's Bureau's popular child-rearing pamphlet, still advised women to hire a wet nurse if they were not going to breast-feed—but mothers had been encouraged to breast-feed for decades and many now did. (A few European countries even subsidized women to stay home and nurse.) The rest simply used formula. Sanitation had dramatically improved—milk was now being pasteurized—and artificial foods were safer and simpler and far more popular. Many mothers nursed for several months and then switched to formula, which was marketed relentlessly: by the 1920s, a new mother in a major American city could expect to be visited by dozens of infant food salesmen, pitching a bewildering array of proprietary products, all containing an equally bewildering diversity of ingredients. In 1934, an article in *Parents* magazine began, "Once you are at home and the baby safely tucked into his little crib, you will be confronted, almost immediately, with the practical necessity of warming and giving some of the formula which the hospital has so kindly provided your baby."

In the new preemie wards of hospitals, though, wet nursing of a sort survived: hospitals stockpiled breast milk in human milk banks—wet nurses now filled up a bottle, not a baby. For a brief period, human milk looked like it was on the verge of becoming just another commercial food product, a commodity as divorced from its origins as a bottle of whole milk. As an ingredient in the food industrial complex, it was manipulated in ways unthinkable today: "Mother's Milk: Babies Now to Get Nature's Food in Cans" read a 1934 *Newsweek* headline.

Human milk banks still exist today. They dwindled in number after World War II, and even after a recent resurgence, there are now only ten scattered throughout the country. With a few posh exceptions—a staffing agency in Beverly Hills advertises actual wet nurses—human milk banks are the only inheritors of the ghostly history of women feeding other women's children. But the milk banks have scrubbed it of ghosts: they do not pay for milk. In defiance of centuries of unsentimental calculations, breast milk is no longer something to be bought or sold. We have decided, at this late date, that it is too valuable to be worth anything at all.

This Is Not a Breast

Wet nursing may seem strange, but there's nothing novel about the concept. It's just a baby at the breast. In a perfect world—a healthy mother, copious milk—it usually works wonderfully. If we find it strange, it is only because of the small matter of whose baby and whose breast.

Artificial feeding, on the other hand, is something that no longer seems strange at all: it is how the majority of infants are fed today. But the history of artificial feeding is in fact far stranger than that of wet nursing. In a perfect world, it usually worked disastrously.

Amid all the high-pitched arguments over the relative benefits of exclusive breast-feeding and the politics of infant formula, it is easy to forget how recently the machinery of artificial feeding stopped malfunctioning. Only in the twentieth century—and not nearly everywhere, not even today—have infants fed without breast milk been more or less assured of survival. Clean water, better formula, rudimentary knowledge of how diseases are spread: only after all these developments did babies without breast milk finally have the odds in their favor.

Dry nursing, the term for feeding infants anything other than breast milk, has rarely been a first choice. But it has never actually been rare, and although many infants died swiftly from artificial foods, many survived, too. The idea that all babies deprived of mother's milk were condemned to death is a myth.

Like wet nursing, dry nursing has ancient origins, although since many infants were just given premasticated food directly from the mouth of an adult, little evidence of it has survived. But archeological digs in Athens and Cyprus have uncovered some infant feeding pots, elegant teapot-like pitchers with shyly puckered spouts. They are beautiful objects with a tragic design flaw: they were almost impossible to thoroughly clean. Germs were the cause of infant mortality as much as poor nutrition, if not more, and the fateful design of these ancient pitchers would be imitated for millennia, inadvertently condemning many infants to death. In a painful irony, the best preserved of the pitchers were found in the graves of infants: to provide comfort in the world to come, the children had been buried with the very objects responsible for their burial.

What was inside those pitchers? Just about everything imaginable. The universal response to the problem of what to feed an infant besides breast milk seems to have been something meticulous like, *Let's see if this works.* Everything looked like it just might be infant food, and over the centuries, infants have been fed just about everything. In much of Europe, this infinite array of dishes was eventually boiled down to something that must have looked a lot like sludge: pap, the most common baby food for centuries, was a mixture of milk or water with flour or mushed bread. Add broth and butter for panada, wine or beer for caudle. (The alcoholic beverages, even in substantial quantities, might have been less dangerous than unboiled water.) As if to compensate for the nutritional deficiencies, infants were stuffed beyond all reason: in a single feeding, as much as two quarts of pap at a time were poured into a baby's mouth, funneled through pap boats, shallow feeding dishes with broad front lips. Babies were fed until they threw up, and then they were promptly fed again; only then were they considered full. Children thrived when they threw up freely, an English physician wrote in the late eighteenth century, reflecting the medical consensus of the time.

The industry of infant feeding was not a hotbed of innovation. It was not until the rubber nipple, invented in 1845, that a new technology would revolutionize how babies were fed. The rubber nipple was an almost high-tech leap over the most successful technology on the market before it: the dried cow's teat.

Thanks in part to the convenience of the rubber nipple, more and more mothers regarded artificial foods as a sensible choice. In

England at the time of the industrial revolution, artificial feeding became popular with all social classes: the upper class out of fashion, the lower class out of necessity. Women who worked in the new mills were given only a few days off after birth, the sort of schedule that made breast-feeding impossible. They were forced to leave their newborns behind. The nannies fed the babies pap. Mortality rates soared.

But artificial feeding was not always a last resort or an elite fad. In wide stretches of northern Europe—southern Germany, northern Italy, much of Austria, Scandinavia, and Russia—dry nursing was the rule, not the exception: it was just how you fed your baby. The historical roots are vague, but the practice was widespread during at least the seventeenth and eighteenth centuries; it may have dated back to ancient times. In the far north, including Iceland and Russia, infants were raised strictly on animal milk; farther south, on pap, or meat soups and mushed bread if necessary. Breast-feeding was treated not as a good backup option, but as something regarded with social shame and even horror. This repugnance was "certainly a deeply-ingrained social custom," according to historian Valerie Fildes, "and any woman who attempted breast-feeding [in these areas] was reviled by women in general, as can be seen from the following example from the Oberbayern district of Bavaria: 'A woman who came from northern Germany and wanted according to the customs of her homeland to nurse her infant herself was openly called swinish and filthy by the local women. Her husband threatened he would no longer eat anything she prepared, if she did not give up this disgusting habit.'"

Fildes, who did seminal research on this subject, cannot explain why these regions rejected nursing. Records are scarce. There are only theories: it might be because social norms placed women in the field, not at home with a newborn, or because garments were so tight-fitting that they made nursing extremely difficult. It might be because in rural areas mothers were gone for much of the summer, gathering food for the rest of the year. It might have started with men forbidding their wives from nursing, fearing "that the breast-feeding will make the women ugly." But none of these explanations fully account for the visceral reaction—the open horror—in the account of the "swinish and filthy" mother in Bavaria.

It's frustrating not to know more. Even when considered against the broad backdrop of boneheaded child-rearing customs, dry nursing seems among the least adaptive, most counterproductive traditions conceivable. It's the sort of child-rearing advice Malthus would have given. In comparison, concluding that colostrum is a toxic, highly poisonous substance seems like a model of reasoned inquiry.

Cultures that relied on artificial feeding may have devised better methods for it, like routinely boiling water. They did tend to have cold, dry climates, in which milk took longer to spoil and babies were less vulnerable to gastrointestinal diseases. In the warmer climates of southern Europe, any society that relied on artificial feeding would surely have suffered truly unsustainable infant mortality rates. Nevertheless, even in northern Europe, the consequences were predictable. In Iceland, after at most a week of breast-feeding, babies were fed either milk, cream, or fish mixed with cream. The diet, according to a visiting doctor in the eighteenth century, killed most of the babies it was supposed to be saving. In Bavaria, a region where different provinces had very different attitudes toward breast-feeding, infants born in areas where nursing was common were more than twice as likely to live to their first birthday. In Sweden, babies born in parishes where there was little breast-feeding were three times more likely to die than babies born in nearby parishes where most women nursed.

But if more babies died in these societies, more babies were also being born. Women who are not nursing become fertile shortly after they give birth. In a society in which no one is nurs-

ing, birthrates are extremely high. This is the perverse physiological logic of dry nursing: it ensured that the population would not
be decimated by it.

We have now come to the livestock. Yes: the livestock.

According to the Greek myths, Zeus was suckled by a wet
nurse named Amalthea. In one very significant way, Amalthea
was not a conventional wet nurse. It seems unlikely that Zeus
really threw thunderbolts, or hung one of his girlfriends upside
down from the firmament, but being nursed from birth by a goat
is different: it is completely credible.

For millennia—in all likelihood, for as long as animals have
been domesticated—humans have sucked at the mammary glands
of other species. Goats, sheep, donkeys, cows, mares: there are
stories even before Zeus of infants—orphaned, abandoned, syphi-
litic, simply undernourished—who were nursed by animals.

Goats were best; asses were good, but pricey. Mares were too
dangerous, though, and the udders of cows far too large. Whatever the species, when families were without options, animals
were always worth trying. From his home in the French countryside, Montaigne, writing in the late sixteenth century, observed
that "it is common around here to see women of the village, when
they cannot feed the children at their breast, call the goats to the
rescue." But cross-species nursing was hardly confined to the rural poor. In the early nineteenth century, the German author Con-
rad Zwierlein even inspired a brief continental vogue for hircine
nursing with the book *The Goat as the Best and Most Agreeable Wet-
Nurse*. The reasons were physiological and psychological: "The
shape and size of the [goat's] teats," according to *A Theoretical and
Practical Treatise on Midwifery*, published in France a few decades
later, "which are easily seized by the child, the abundance and
quality of its milk, the docility of the animal, the ease with which
it is trained to give suck to the child, and the attachment which it

is capable of forming for it, are sufficient reasons for the preference." The *Treatise* cautions that "the petulance and impatience of the animal expose it to frequent accident, but after a time the goat comes of its own accord to give it suck. The infant should be laid in a low cradle placed upon the floor." In the 1860 book *Infant Feeding and Its Influence on Life*, the English physician Charles Henry Felix Routh described an infant who'd been nursed by a goat:

> Mrs. P, being much annoyed with a wet nurse whom she had in her service, was at last compelled to part with her. Being aware that a friend in the neighborhood was possessed of a goat, it was asked for to supply the wet nurse's place. The child took to the goat at once, and the poor thing after a short time would run up the stairs when she was loosed for the purpose of supplying a meal, and the child would suck the goat like a young kid . . . The little boy did at any rate exceedingly well under this treatment.

Some hospitals employed goats to suckle any babies with congenital syphilis, a disease that could be transmitted by nursing. At a hospital in Aix en Provence in the early nineteenth century, a visiting doctor reported that the goats not only recognized the infants under their care but went to nurse them on their own initiative: the goats were said to stand over the cradles. A century later, a visitor to a different French hospital reported that "goats came to know the babies so well that each would run to her nursling when she heard its cries."

The primary problem with infants suckling at animals was not any nutritional deficiency—given that dry feeding was often a death sentence, a goat amounted to a reprieve—but the persistent belief that babies inherited the worst qualities of their wet nurses. This was frightening enough with redheaded wet nurses; with another species altogether, it was terrifying. Infants nursed by animals were said to be "fierce and not like men." Such anxieties kept this sort of feeding from becoming far more common. In an era when the death of a child was an unavoidable part of life for many families, what a child might turn into seemed to have been a greater fear than the child not surviving at all.

Against All Odds

Historical infant mortality rates are shocking: up until at least the beginning of the nineteenth century, and possibly much later, it is thought that 30 to 40 percent of babies frequently died before they turned a year old. These days, 10 percent is shocking. (The United States averages around .6 percent and it lags behind many other developed countries.)

But when you realize how infants were treated, it is less surprising that so many died. What's surprising is how many *survived*.

Colostrum is a case study in how the ideologies of infant feeding have failed infants. But it isn't exceptional. The history of how babies were fed is a thick volume of erroneous convictions that didn't collapse until long after they started to sound hollow. It's hard to imagine that there are other mammals who have been deliberately subjected to a worse, less sustainable diet than the human newborn. We need culture to survive; we can't figure out how to breast-feed without it, after all. (Even gorillas can't!) But culture can kill.

Our physiology seems to anticipate this sort of treatment: babies emerge from the birth canal padded with fat, far chubbier than other mammals. Newborns weigh in at more than 10 percent body fat; pigs, who will soon have far more than that, are born with just 2 percent. After birth, newborns are a bit like hikers who have taken a wrong turn and found themselves in a strange, unmapped land with empty knapsacks. They draw down their fat to survive.

Historically, there was a lot to survive. Babies had to survive being fed, not breast milk, not cow's milk, but a little cow's milk, a lot of water, and a slew of ingredients meant to *simulate* milk: grilled onions and caramel and marigold petals for the color; plaster and whitewash and clay for the viscosity; and for the cream, emulsions of animal brains. They had to survive being sent away to wet nurses without any immunities, powerless against any infection. They had to survive being stepped on by goats.

But newborns have proved to be remarkably sturdy. From the trash heap of ways babies have been fed, that's the unavoidable conclusion: for a fragile creature, born almost a year too soon, the human infant has managed to put up with an astonishing amount.

Of course, there are some insults infants can't absorb. The problem is, we don't always know what they are, and they aren't necessarily what we'd assume they are. A society that rejects breast-feeding is dunderheaded but not doomed. An infant suckled by a goat might do surprisingly well (and he might avoid toxic milk and water). With some common sense, wet nursing could work out just fine. The worst insults were often delivered by doctors; they weren't intended to be insults at all. For a very long time, the hardest things for infants to survive were the cures.

The history of babies is much messier than we might imagine. There's a tendency to assume that the past is purer than the present—that between formula and full-time day care, we're tampering with the fundamental, timeless conditions for child rearing. But when Rousseau, for instance, called for the mothers of France to take their infants unto their breasts—when he said nursing was the wellspring of a moral society—he wasn't calling for a return to the past: he was *inventing* a past. There are no timeless conditions for child rearing, even when it comes to an activity as fundamental as feeding. Deviation is the norm.

4

Tasty, Tasty Thumb

To review, a very brief history of what babies have sucked: their mother's breasts, the breasts of other women, the breasts of other species altogether, the crude nipple-shaped ends of germ-festering food pots, the pert silicone of modern bottles.

What's missing is what's caused the most consternation of all: their bodies, themselves.

No baby would miss this. It's been estimated that 95 percent of newborns suck their thumbs and fingers. The majority of newborns spend more time sucking for no apparent reason than they do nursing. The medical term for this is non-nutritive sucking—any sucking that doesn't fill you up. It's a category that includes thumbs, fingers, pacifiers, blankets, stuffed sheep, a dog's tail, a grandfather's nose. (For simplicity's sake, I'll use thumb-sucking as a catch-all term, except when we get to pacifiers.)

We know, by birth, exactly what to do. Slow a newborn down to frame-by-frame speed: as his hand approaches his face, his mouth, with perfect timing, opens to welcome it. We are programmed to suck our thumbs and fingers: it is what's called a "highly organized action," a motion that our bodies are designed to make. It seems part of our mammalian inheritance. Our primate relatives routinely suck their thumbs. Kittens suck at length for no milk at all; rats do too.

In the last century, we have been at war with this bit of our mammalian inheritance. Since the late nineteenth century, the grown-ups—pediatricians, psychologists, parents—have tried to sever infants from the motions their bodies have insisted on making. Thumb-sucking and later pacifiers have been called the roots of masturbation, the harbingers of lifelong depravity and immorality. They have been linked to horrific facial malformations and crippling personality disorders. They have driven parents to stuff the hands of their children into iron mittens, to paint them with noxious solutions.

These days, we have largely stopped fretting about thumb-sucking, but feelings about pacifiers still run strong and deep, and they rub up against this weird, forgotten history. Our concerns today are more medicalized—we are worried about nipple confusion rather than moral depravity—but they are often no more rational. It shouldn't be a surprise that we're confused without knowing why. After a century of hysteria, there's little precedent for clarity on the subject of babies sucking.

In retrospect. A lot of books about babies could be shrunk to nothing but those two words: in retrospect, washing cloth diapers together did not make the relationship stronger. In retrospect, the attachment parenting model would have worked better with more detachment. In retrospect, generic condoms are a poor place for cost savings.

My first *in retrospect* isn't about sex, or sanity, or sex and sanity. It's about something smaller, available in either plastic or silicone, with federally mandated ventilation holes: in retrospect, I should have let Anya buy the pacifiers.

When Isaiah was a month old, I found myself standing in front of the pacifier display of the baby department with a shopping list that had *pacifiers* underlined and in all caps. What happened next should have been mundane and unmemorable. But as I stood there, newly aware of the fact that pacifiers now come in fake-crystal-embedded bling, I realized I was about to walk out of the aisle empty-handed: *I was afraid to purchase a pacifier.*

It was an embarrassing, vaguely emasculating feeling for a new father to have. I was afraid to purchase a pacifier! How was I going to deal with the monsters under the bed?

I knew I disliked pacifiers. I'd inherited the feeling from my mother, who thought they were bottle-stoppers for babies. But I'd thought my distaste was only that. Refusing to buy a pacifier for Isaiah amounted to an act of self-loathing: it was going to make my life more miserable. (He was a month old: it was already miserable.) It was also going to make Anya's life more miserable, and although being unable to explain my own indecision to myself was bad enough, being unable to explain it to her would be far worse. I needed some sort of story. I could have told her that none of the brands met our acronym-conscious criteria. I could have said that the only pacifiers left had PIMP spelled out in fake diamonds. I could have said that a horde of angry, orally fixated toddlers stormed into the store and hoisted the entire pacifier display over their heads and out the door. These stories would have all been credible. But I could not have said that I had forgotten. That would have been incredible.

In the end, I settled for incomprehension:

Me: Icouldn'tdoit.

Her: Mum. Mum. Mum. Stop mumbling. And where are the pacifiers?

Me: Icouldn'tbuythem. Idon'tknowwhy. Ihatethem.

Her: Really? But they're so cute.

Me: [*Jesus. What am I doing with this woman?*]

She went back to the store. Isaiah liked his pacifier (blue and green, no bling). He went easier on our ears. Life got better, at least around the margins.

But I still needed a story. I needed to know why I'd freaked out.

"Fools, Fools, Fools, You Know Not What You Do!"

Before pacifiers came thumbs. Thumb-sucking, like nose picking or groin scratching, is left out of the history books. No ancient scribes were ever assigned to take detailed notes on it. No one finds it preserved in any archaeological digs. It vanishes along with all the other nervous twitches of history.

The few references to the habit in the historical record are unremarkable: it is called soothing, comforting, normal. Then, after millennia of apparent quiet, there's an uproar.

In 1879, a disturbing illustration of a "6 year old thumb pleasure-sucker with active assistance" appeared in a German medical journal: a girl contentedly sucking her thumb while—very graphically—slipping her hand under her dress. The journal article—"The sucking of the fingers, lips, etc., by children (pleasure-sucking)," by a Hungarian pediatrician named S. Lindner—was the first "scientific study . . . devoted to pleasure sucking." Lindner concluded not only that every child was driven to suck compulsively, but that "pleasure sucking" was the cause of chronic masturbation. In a sort of Kama Sutra of infantile sucking, he described how a child "could suck not only the thumb but any other body part, or it could combine sucking and masturbatory manipulations of the body by the hands or feet." His evidence was a study of sixty-nine children, and as Lindner himself admitted, only four of those studied sucked with "the active assistance of the genitals." Nonetheless, he took the small number to be proof of his hypothesis:

You might want to counter here, certainly not without jus-
tification, that four masturbating children . . . hardly permit
me to count pleasure-sucking among the many causes for
masturbation . . . However, Gentlemen!, I would like to point
out that many of my pleasure-suckers came from circles where I
did not consider it socially proper to inquire or investigate more
deeply. It is possible, indeed probable, that my results would have
been of wider implication had there not been such obstacles.

Lindner wrote just before the expertification of child rearing
and you can tell: he was all too transparent about his fallacious
argument. (The experts who followed him would learn to be
more opaque.) Unbelievably, though, his logic carried the day.
Lindner's journal article is the study that launched a thousand
parental nightmares. It's the urtext of orality—it would inspire
Freud several decades later—and by the turn of the century, Lindner's leapfrog of a conclusion was widely accepted. Twenty years
after Lindner's paper, the most prominent pediatrician of the day,
Luther Emmett Holt, wrote in his definitive pediatrics textbook
that "the most pernicious result of sucking is its tendency to develop the habit of masturbation." As support, Holt referred to
Lindner: the behavior of four German children, recorded by Lindner and laundered by repeated reference in other pediatric papers
and textbooks, had been transformed into unimpeachable evidence that thumb-sucking led inexorably to masturbation.

It was all enough to make *Good Housekeeping* hyperventilate.
From a 1908 column:

I watched a young mother the other day sowing seeds of
trouble, mortification, exasperation, worn nerves, for herself;
pain, rebelliousness, possibly permanent disfigurement, for her
child. It filled my soul with wrath. What was she doing? Encouraging a habit; "cute," she called it, but one which she will

some day know for what it is—*pernicious* to the last degree. This is the thumb-sucking habit.

I have heard all the arguments of doting parents. "It is such a comfort to him;" "It keeps him so quiet and happy;" "It would be cruel to deprive him of so harmless a pleasure;" "He'll drop it by and by." Every time I am tempted to cry aloud: "Fools, fools, fools, you know not what you do!" . . . when supposedly intelligent parents "let um 'ittle tootsy-wootsy suck um fum if um wants to," and allow the establishment of other foolish or harmful habits just because baby looks "cute," I get up and leave, lest in my wrath I speak my mind. There are great evils to guard the child against, but there are lesser evils for which the years will surely exact penance, and thumb-sucking is one of them.

Pity the poor mother who read *that* and then discovered her son with his thumb in his mouth. But there was little pity at the time for mothers who were not prepared to be ruthless. "It may be that there are mothers in the world who are so weak and indulgent that they can not break up a harmful practice lest the dear child be caused some present discomfort," the American dentist Samuel Hopkins wrote of thumb-sucking. "Such women are unconsciously wicked." For Hopkins, "So hideous is the deformity caused by this habit, that it seems incredible that it should be necessary even to call attention to it, much less to urge that action be taken to put a stop to the evil."

In a few short decades, thumb-sucking had been transformed from a harmless habit into an evil. Pediatricians, who'd hardly given a thought to thumb-sucking before, had helped make it so. In the late 1800s, the specialty of pediatrics wasn't even a specialty. Babies were regarded as either the responsibility of midwives and mothers, in which case they required no doctor, or as miniature adults, in which case they simply needed the same sort of doctor as everyone else.

Pediatricians needed a condition of their own, a disease that would justify a child-only specialty. Ideally, it would be a common condition that mothers and midwives had overlooked, or even encouraged, and that was found only in infants and children. Thumb-sucking was made to order: infantile and either ignored or abetted. It was a gift from the disciplinary heavens.

"Pediatric textbooks came to classify the habit as a neurological disease which could only really be understood and managed by specialist pediatricians," the medical historian Jonathan Gillis says. How could a mere mother hope to treat a neurological disease? Never mind that neurology textbooks failed to even mention thumb-sucking. For pediatricians, it was formally classified as a neurological disorder. The very fact that mothers and neurologists had dismissed the habit was precisely *why* the discipline of pediatrics was necessary: only a trained pediatrician could recognize the true danger hidden behind a seemingly innocent habit like thumb-sucking.

It was not always thus. A charming poem published in *Ladies' Home Journal* in 1891—when the clouds of thumb-sucking were still gathering—suggests how differently the habit was once seen:

> *If a babe sucks his thumb*
> *'Tis an ease to his gum:*
> *A comfort, a boon, a calmer of grief*
> *A friend in his need affording relief*
> *A solace, a good, a soother of pain;*
> *A composer to sleep, a charm and a gain . . .*

Within a decade, this bit of verse—kind-hearted, empathetic, generous—would be an anachronism. At the turn of the twentieth century, the child rearing of the Victorian era—which, despite its reputation, was often indulgent and sentimental—was at odds with a new, harsher era. The Victorian model was too sentimental, too loosey-goosey, especially in contrast to the new, supposedly scientific school of child rearing—stricter and highly regimented—that was on the horizon. The claims made for this new science of child care extended far beyond the child. The fundamental problems of society—poverty, immorality, class conflict—were reformulated as problems of lousy parenting. If the errors of child rearing were corrected, the social problems would fix themselves.

Mother no longer knew best; she now knew almost nothing. Expert-devised schedules mandated rigid timing for when all infants ate and slept, and mothers deviated from those timetables at their peril. Undesirable habits were stamped out without pity. Even crying—yet another undesirable habit—was never to be indulged. "The child," as the historian Steven Mintz has phrased it, "was seen as a small animal with fearsome appetites, who had to be broken in."

No appetite was as fearsome as a baby's hunger for his thumb, and in the child-care manuals and pediatric textbooks of the early twentieth century, the new scientific ideology went to war against the indolent habit of thumb-sucking. Pediatricians, psychologists, and parents all tried seemingly everything—they threw up every obstacle they could think of—only to have the child *keep sucking his thumb*. They had been outwitted by a baby.

And so, as babies wiggled and squiggled their way to their thumbs, an arms race of anti-sucking devices ensued. Children woke to find their nightgown sleeves pinned to the bed, or their arms tied or taped to their sides. To prevent babies from bending their arms, splints or bandages stiffened with tongue depressors were strapped around their elbows. Many books provided sewing pattern–like instructions so mothers could make their own contraptions. Ingenious anti-sucking devices were invented. A 1923 issue of *Popular Science* featured a metal thumb band with dangling metal links. When the child sucks his thumb, "the metal parts come in contact with the tongue and the roof of the mouth. The result is unpleasant, but not painful."

To disseminate the new modern wisdom of child care, in 1914 the United States Children's Bureau began to publish a bulletin of child rearing advice, periodically updated, called *Infant Care*. Instantly popular—some 1.5 million copies of the first edition were distributed—the bulletin was a state-of-the-science account of the New American Baby, as envisioned by the experts, and its treatment of thumb-sucking was uncompromising. Breaking the bad habit of thumb-sucking would not be easy, the authors of the 1914 bulletin counseled mothers; the baby's hands would have to be continually covered. Apparently worried more about *over*compliance than mere compliance, it was noted that "the baby's hands should be set free now and then, especially if he is old enough to use his hands for his toys, and at meal times."

The bulletin was no less severe with pacifiers. Like Cato the Elder declaring war on Carthage, the authors of *Infant Care* insisted that "the pacifier, of whatever variety, must be destroyed." Pacifiers were particularly infuriating to the experts, on both sides of the Atlantic, who were incredulous that mothers would *introduce* an object designed for sucking. "Remember that a baby that has a dummy is like a tiger that has tasted blood," warned an English health pamphlet, using the British term for a pacifier. A popular English child-care book of the time described the typical pacifier-sucking child as "ricketty, pale, pasty, soft, wanting in bone and muscle, feeble, nervous, timid."

The obstinate habit of thumb-sucking was soon pitted against an even more determined adversary: behaviorism, the idea that every aspect of behavior is learned. The 1920s were the high-water mark of the behaviorist ideal in American child rearing, exemplified by the skyrocketing success of the psychologist John Watson. "Children," Watson insisted in his book on the proper molding of children, *Psychological Care of Infant and Child*, "are made not born." A man who seemed to know neither self-doubt nor subtlety, Watson's bolder claims for behaviorism are infamous: "The world would be considerably better off," he wrote, "if we were to stop having children for twenty years (except for experimental purposes) and were then to start again with enough facts to do the job with some degree of skill and accuracy." He was sure of his own skill and accuracy: "Give me a dozen healthy infants, well-formed, and my own specified world to bring them up in and I'll guarantee to take any one at random and train him to become any type of specialist I might select—doctor, lawyer, artist, merchant-chief and, yes, even beggar-man and thief, regardless of his talents, penchants, tendencies, abilities, vocations, and race of his ancestors."

For Watson—a man for whom everything was training—the persistence of thumb-sucking must have been humiliating. If it continues past early infancy, Watson told his readers, they must "take more drastic steps to break the habit. Sew loose, white cotton flannel mitts with no finger or thumb divisions to the sleeves of the night gown and on all the day dresses, and leave them on for two weeks or more—day and night." He continues, censoriously, "So many mothers leave them on only at night." But even the model behaviorist admits he cannot get any method to truly work. Even physical punishment—"sharply rapping the finger with a pencil"—fails: it is only "beautifully effective while the experimenter is around." The infant had learned—Watson should have been proud—that the experimenter eventually goes away. The thumb never leaves.

The Past Is Another Parent

The history of child rearing is so gothic, you can lose sight of how strange the present is, too.

A few years ago, a friend's young baby—let's call her Mary—

had become a little too attached to her pacifier. The two were too close: she needed the pacifier at all times; she was lost without it. It wasn't Romeo and Juliet, at least not yet, but her parents weren't happy with the relationship.

Then Mary's grandmother arrived. For an entire weekend, the grandmother *held Mary's thumb in her mouth*. It hardly left her mouth the whole weekend, the story goes. And by the end of the weekend, Mary had forsaken her pacifier for her thumb. It's a very traditional story, really: the child chooses her partner; the family disagrees; the matriarch arranges a different match; they grow to love each other. The end.

The reason I like this story so much is that it seems to belong in a child-care manual from a century ago. The baby isn't trusted with her own hands; they must be controlled for her. There's only a minor difference, but it makes all the difference. She isn't being prevented from sucking her thumb. She's being *made* to suck her thumb. A lot has changed since Lindner.

And yet. Our anxieties about children are always locked in time, and outside of it, they stop making sense. This is never more true than during the fever dream of new parenthood. At the time we heard about Mary, her thumb, and her grandmother, I hated Isaiah's pacifier. I can still see him, in his stroller in the park as we listen to the story, sucking contentedly on silicone. I dearly, desperately wanted him to suck his thumb. This story of successfully imposed thumb-sucking was inspiring.

It isn't anymore. It seems, well, less than inspiring. After only a couple of years, I am already alienated from my own reactions. I am locked out of my own parental past.

The measures once taken against thumb-sucking seem extreme, even cruel, but they were not irrational. Experts and parents weren't afraid of behavior in babyhood; they were afraid of behavior in adulthood. They believed the child was permanently marked by the habits of infancy. If bad habits weren't stamped out, they'd sink in. You'd end up with a compulsively masturbating middle-aged son or daughter. Faced with that future, what was a loving parent to do?

The Orgasmic Infant

Today, of course, when we think about compulsive sucking as a sexual act, we don't think of Lindner, or Luther Emmett Holt, or John Watson. We think of Sigmund Freud, who sketched out the concept of orality in *Three Essays on the Theory of Sexuality,* his seminal work on human sexuality and the source of ideas like the Oedipal complex, castration anxiety, and penis envy. In 1905, the book went off like a bomb, and the fragments scattered over the rest of the century, lodging in just about every sphere of modern intellectual life.

But by the time *Three Essays* was published, sucking's sexual nature was, at least among pediatricians, already a matter of medical record. In fact, Freud was inspired by Lindner's work—he cites him in *Three Essays.* Freud was original not because he called sucking sexual, but because he redefined what sexual meant: rather than calling childhood sexuality an aberration, Freud insisted it was normal, and that adult sexuality was its natural continuation.

Freud's theories of child sexuality may not have been as unique as he claimed, but they were still shocking. His discussion of infantile sexuality was intimidating, even terrifying, but his fundamental point was supposed to be consoling: he was arguing that sexual behavior in children is *not* perverse or sinful. But to make that point, Freud had to describe what he saw as sexual behavior in children, a description his readers could only see, no matter how Freud glossed over it, as perverse and sinful.

For all the analysis it has borne, Freud's theory of sucking is not, in the end, especially complicated: he thought all sucking after a baby's hunger had been satisfied was sensual, that this was how human beings discovered sensuality, and that this discovery unlocked the sex drive. That's why thumb-sucking, pointless to any adult, was so obviously pleasurable to the infant. It was, in a word, orgasmic:

> No one who has seen a baby sinking back satiated from the breast and falling asleep with flushed cheeks and a blissful smile can escape the reflection that this picture persists as a prototype of the expression of sexual satisfaction in later life.

The psychiatrist Carl Jung, who'd just broken with Freud, thought otherwise. "Obtaining pleasure is by no means identical with sexuality," Jung wrote at the same time. "We deceive ourselves if we think that in the suckling both instincts exist side by side, for then we project into the psyche of the child the facts taken from the psychology of adults." Jung's argument has won the century: these days no one compares the bliss of a newborn sucking to late-night Cinemax.

Given the scandalous ring of phrases like "sexual satisfaction," it is easy to forget that Freud did not think sucking was deviant; he thought it was normal, even necessary. But it wasn't physiologically important. It was *psychologically* important: infantile sucking was the cornerstone of a well-adjusted personality. Any good mother should make sure her infant had a healthy amount of time to suck. Unfortunately, it was never clear how much a healthy amount was. The proper quantity of sucking was subject to a near-Heisenbergian level of uncertainty: you couldn't know what the right amount of sucking was while the child was sucking. A mother only knew what she'd done—if she'd allowed too much or too little or just enough—when she saw the consequences.

These were grave. As Freud wrote in *Three Essays,* children who suck excessively will grow up to "become epicures in kissing, will be inclined to perverse kissing, or, if males, will have a powerful motive for kissing and smoking. If, however, repression ensues, they will feel disgust at food and will produce hysterical vomiting." It's hard not to feel that too much sucking would be a far preferable fate: an adulthood of epicurean kissing sounds, at the very least, less messy than one of hysterical vomiting. The psychoanalyst Karl Abraham, whom Freud called his most brilliant pupil, would later elaborate on the type of whiny, needy people who hadn't sucked enough in infancy: "The manner in which they put forward their wishes has something in the nature of persistent sucking about it; they are as little to be put off by hard facts as by reasonable arguments, but continue to plead and insist. One might almost say that they 'cling like leeches' to other people." In some cases, Abraham added, "their behavior has an element of cruelty in it as well, which makes them something like vampires to other people."

Because of Freud's sheer strangeness, his work took decades to be absorbed in the child-rearing manuals: it was not until the 1940s that the language of Freudianism seeped into the advice of the experts. It took so long, at least in part, because Freud's analysis was never practical: he had no advice to offer mothers; he had only consequences.

But there was another reason for the lag. People were only willing to follow Freud so far: even if parents took his argument seriously, they had no intention of taking his *conclusions* seriously. A few people were willing to believe that infantile sexuality was real; far fewer were willing to believe it was harmless. A special article on thumb-sucking in *Pediatrics*, published in 1956—a half century after *Three Essays*—saw the self-contradictory nature of the problem clearly: "Since open expression of sexual impulses is taboo in our culture it became clear, as Freud became more and more popular, that a baby certainly should not suck his thumb."

Writing at the same time, the psychologist Martha Wolfenstein, looking back at the previous forty years of *Infant Care* manuals, noted that infantile sexuality was no longer being ignored. It was just drained of any significance. The gentler tone toward masturbation in the manuals, Wolfenstein observed, "is accompanied by an increasingly diluted version of the activity. From expressing an urgent and dangerous impulse of the child, masturbation becomes an act about which the child has no feelings and which is only inexplicably embarrassing to the mother." Freud had suffered the greatest insult a radical thinker can bear: he had been domesticated.

It is easy to quibble with Freud's hot-hot-hot interpretation of sucking, or to mock his red-light district account of infancy. But Freud was asking something about sucking that no one else had yet. He was asking *why*.

Before Freud, no one had shown much interest in the question. The few doctors who'd stopped to consider thumb-sucking had shaken their heads and walked away, bewildered. William Preyer, the founder of child psychology, could hardly hazard an explanation for what he termed "useless sucking." In 1882, Preyer concluded because the milk in the breast was invisible, the infant must think there might be milk in the fingers, too. But Preyer acknowledged that explanation didn't make much sense. After all, why didn't the baby stop sucking after he realized his mistake?

Indeed, this simple, stubborn question—why did babies insist on sucking for *no reason whatsoever*—would become a major debate in psychology. A book surveying important problems in the discipline, published in 1962—approaching a century after Lindner—included a lengthy chapter on thumb-sucking. The habit, the authors noted, "has attracted much attention from mothers, orthodontists, pediatricians, psychologists, and nearly every other group interested in the development and welfare of children." They went on:

Sucking, and particularly thumb-sucking, is a phenomenon that excites intense and widespread interest among both laymen and scientists . . . It is hard to believe that there has ever been so intense and widespread an interest as during the last forty years. No mothers' manual is complete without a discussion of the subject; no pediatric handbook fails to mention it; preoccupation with the question is almost an occupational disease among orthodontists; and personality theorists appear to feel obliged either to assert or deny the importance of early oral experience.

The research piled up. Sometimes it was weird and brilliant, and sometimes it was just weird. To wit, the article "Toe-Sucking in Baboons: A Consideration of Some of the Factors Responsible for This Habit," with its sublime opening sentence: "Although the writer has witnessed foot-sucking on numerous occasions in two young honey bears the present paper is devoted mainly to a record of thumb- and toe-sucking in the baboon."

It was not all baboons and honey bears. There were other primates, and cocker spaniels, and piglets, and puppies, and rigorous investigations into the tail-sucking habits of dairy and beef calves. The experiments were conducted by the warring camps of the time, the psychoanalysts and the learning theorists. The psychoanalysts wanted to prove that there was an inborn need to suck—that if a newborn isn't given sufficient opportunity to satisfy his need to suck, he'll seek it out on his own. The learning theorists wanted to prove that sucking was, well, learned—that it was a habit, not an instinct. If babies didn't learn it, they'd never have it.

Both camps had come to opposite conclusions; both camps conducted experiments that proved their theory; and both replicated their findings with numerous other experiments. It was an argument seemingly without end.

But outside of academia, the argument was already ending.

No longer an all-caps crisis, thumb-sucking became a lowercase, optional problem—something to be worried about if you could work up the energy.

The shift happened in a hurry: the 1938 edition of *Infant Care* recommended stiff cuffs to prevent thumb-sucking; the 1942 edition recommended *against* using restraints like stiff cuffs to prevent thumb-sucking. By 1951, the publication was so sympathetic to the infant that it sounded genuinely bewildered by all the hubbub: "Why do so many of us have this strong feeling against what is so perfectly natural for babies to do? Somehow the idea grew up that it was bad and harmful."

There were still reasons for concern—they'd just changed. Benjamin Spock's books, when they arrived just after World War II, represented a gentler, kinder form of Freudianism: he was worried about how best to nurture a well-adjusted personality, not how to mold character. This new school of thought provided a new explanation for why thumb-sucking persists. In 1942, *Infant Care* told mothers who were worried about the habit to worry more about themselves: "Giving him comfort and satisfaction in other ways, such as a little more attention, a little 'mothering,' will help him to give up sucking." Mechanical devices or bitter ointments only "make [the child] irritable and unhappy. It's easy to see they do nothing toward banishing such causes of sucking as loneliness or boredom."

Thumb-sucking wasn't the cause of serious problems anymore. It was an *indication* of serious problems, a canary that could reveal a hazardous climate for the growing child: "It is clear from numerous pediatric sources that healthy . . . babies suck their thumbs. It is not so clear that *happy* babies do." In this new Freud-tinted age, the problems were "loneliness or boredom" or just unhappiness—problems a loving mother could prevent, rather than problems deeply rooted in the child himself. Thumb-sucking was no longer a sinful act. It was a cry for help.

This link between thumb-sucking and distress seemed solid.

It was anecdotally convincing, at least. But in a journal article modestly titled "Sucking in Infancy," the young T. Berry Brazelton, years from the celebrity he'd attain as "America's pediatrician," argued that this interpretation of thumb-sucking was upside down: it wasn't a marker of unhappiness; it was a cause of happiness. "The importance of this kind of sucking as a source of gratification has often been overlooked," Brazelton wrote. "It seems to be common in the first year for babies to seek extra sucking, and to enjoy it. It does not necessarily appear to be a manifestation of unusual tension or frustration." His evidence was a yearlong study of a large group of infants, which documented that non-nutritive sucking is "common in healthy and contented babies." Brazelton's study would be buttressed by a longer-term study, published in 1971, looking at behavior from birth until age eighteen. It found that the children most likely to suck their thumbs weren't neglected or emotionally deprived. They were the very opposite: they were the most likely to have warm and very loving mothers. The results, the author noted, were "the reverse of what would be theoretically expected."

The idea that enough parental love—enough proper "mothering"—would prevent thumb-sucking was discarded. It was tossed onto a rubbish pile that was overflowing—virtually every explanation for thumb-sucking proposed since Lindner had ended up there. Thumb-sucking hadn't produced a new breed of compulsive, morally degenerate masturbators. It hadn't created a generation of children with repulsive, horribly disfigured faces. Insufficient sucking hadn't turned those children into leech-like parasites; excessive sucking hadn't turned them into mouth-addled addicts. Almost a century's worth of psychological theories—sometimes baroque, often ominous, always utterly certain—had accumulated dust for evidence.

All the while, babies, who are good at keeping secrets, kept sucking.

The Heroin of Babies

On the seemingly unending list of very eccentric convictions adults have had about babies, the various misunderstandings of thumb-sucking must rank near the top. Despite all the complicated theories for thumb-sucking that have flourished, the best explanations today for what's behind the behavior are all rather straightforward.

The simple act of sucking generates a wave of striking physical changes: the heart rate drops; breathing slows. Restless movements quiet down. Babies become less distractible. If asleep, they tend to sleep more soundly, and their sleep states stabilize. Put in the most general terms, thumb-sucking reduces what doctors call, with no historical irony, *arousal*.

In infant brains, sucking runs along the neural routes known as opioid pathways, powerful painkilling circuits—heroin, most notably, uses the same wiring. The journey from Lindner to opioids represents the fundamental shift that's taken place in psychology over the last half century, from theories about the mind—metaphorical, allusive, literary—to neurological data points. This shift has also taken place in *us*. For better or worse, it is how we—possibly doped up on our own neurotransmitter enhancers—naturally think about ourselves now.

Perhaps because it jibes with our expectations, this new explanation of thumb-sucking, on the heels of the much more colorful rationalizations of the past, is vaguely disappointing. It seems dully sensible. What you think thumb-sucking is about is pretty much what thumb-sucking is about. Or at least what *we* think it's about. The last century suggests there's no obvious way to understand thumb-sucking—if there were, there wouldn't have been so much confusion.

Thinking in terms of arousal explains why, amidst all the anti-sucking hysteria, it proved so damn hard to stop infants from sucking: it is the only option they have to reduce arousal—to calm down. They refused to stop sucking because sucking was irreplaceable. The role of arousal also explains why excitement—an extreme of positive emotion—often inspires bouts of vigorous sucking. A study of children who still sucked their thumbs at older ages—when they were between nine and twelve years old—found they relied on their thumbs only when they were excited, not when they were distressed.

It was only when I read this last study that I realized my son was not an inexplicable aberration, previously unknown to science.

Isaiah had never shown the slightest interest in his thumb. A few years after he was born, we were surprised to discover a photo of him at several weeks old with his thumb in his mouth. We'd never remembered him sucking his thumb once.

At least, we never remembered him doing it when he was unhappy. He took the pacifier's name too literally: for pacification, he would accept nothing else. He only used his hands to soothe his *happiness*. When things were too exciting—when the world had revealed itself to be yet more magnificent than he'd imagined—he'd plunge his right hand into his mouth. He was like a ship seeking anchor, and sometimes a single hand was not enough to moor him. I remember a morning when he was teething on his left hand in his crib and he was so amazed by my walking into his room—so amazed that there were people besides him in the world—that he immediately tried to make room for the other hand.

Both hands! The horror! To the poor reader of *Good Housekeeping* a century ago, the child might have seemed ruined already.

You can live in the past, but you can't raise a child there. Today it seems ludicrous to have been so worked up about a habit like thumb-sucking. The anxieties are too far back and too foreign. They're incomprehensible, but to the helpless parents of the time they were all too comprehensible. As a helpless parent of my time,

I cannot read about thumb-sucking without realizing, *Someday my anxieties will be too far back and too foreign, too.* What am I doing now—what am I doing to Isaiah—that will be ridiculed in a hundred years?

Distilled down, the history of thumb-sucking is a cautionary tale so simple it reads like a parable: how, in the not-distant-enough past, people managed to misread the self-soothing response of an infant—an infant who's largely helpless without that response—as something deviant. Incredibly, they misread an almost *universal* response as being deviant: the fact that everyone had the deviant habit failed to convince the experts that it wasn't actually deviant.

The problem with parables is that they have morals, and as a parent, I'm instinctively suspicious of morals—probably because I suspect I'll end up on the wrong side of them. But despite myself, I do think there's at least *a* moral here, and like so many stories about babies, it has to do with power, or better yet, with responsibility. It's about the responsibility that comes from being the only person with any power.

This isn't exactly an esoteric matter. It's a fundamental tension of parenting; every parent struggles with it. (Every book about babies struggles with it, too, although usually not explicitly—it undercuts the confident tone most manuals shoot for.) But the story of thumb-sucking—which is about stopping babies from doing nearly the only thing they *can* do—poses the problem in especially stark form. And what it suggests, I think, is not that parents should only trust themselves or that they should soak every baby book they own in butane. It's something smaller: just because you have all the power doesn't mean you always have to exercise it. And if you do, and it doesn't work, there might be a good reason for it not working.

5

Pacifiers: The Next Menace

If there were a logical point to stop this part of the story, it would be right here. After the 1960s, research into thumb-sucking dried up. Psychoanalysts found other neuroses to dissect. Experts stopped warning against it; parents stopped worrying about it. The subject of babies sucking for no obvious reason seemed to have been exhausted, spent, the casualty of many years of high emotion and low correlation.

Instead, it was reborn as a debate over pacifiers, an argument that was ostensibly more enlightened—with concerns about "nipple confusion" rather than masturbation—but with passions just as intense and often as illogical. Like the reams of papers written on thumb-sucking, the pacifier question—innocuous or insidious—has generated an almost unaccountable amount of research. A physician wrote recently in a review of the breast-feeding literature, "Nobody would have predicted 30 or 50 years ago that at the end of our century the pacifier would be object of a large number of scientific discussions and papers, some of which even had the honour to be published in the most celebrated journals like *The Lancet* and others."

It is not as if the pacifier just arrived on the scene. The pacifier—the dummy, comforter, binky, soother, sucky, titty, silver spoon, coral—goes back almost as far as the thumb. There were ancient clay pacifiers sculpted in the shape of horses and frogs, with holes that could be stuffed with sweeteners. Sucrose is a powerful sedative in infancy, so potent that for several millennia few parents could resist it—the tradition of dipping a pacifier in honey or sugar water only died out in the last century. (It caused the severe dental problems that dentists attributed to pacifiers in general.) For centuries, European children sucked on knotted rags, or "sugar teats," that were tied around foodstuffs and dipped in sweeteners. The combination of sucking and sucrose has an unbeatable punch: during circumcision, a pacifier dipped in sugar water reduces crying more than either alone. Mohels use sweet wine for a reason.

The sugar teat pacifier, along with its many primitive forefathers, did not survive the arrival of the first rubber pacifier: by 1900, rubber pacifiers were standard. This was the historical moment, of course, when thumb-sucking was vilified, accused of all manner of moral and physical harm. But unlike thumb-sucking, which was usually blamed for damaging infants—making them into immoral or mutilated creatures—the pacifier has been accused of actually *killing* them. Germ theory was still in its awkward adolescence, and the pacifier, always on the ground and coated with who knows what, was an obvious target for doctors newly suspicious of "dirt" in general. Convinced it was the cause of staggeringly high infant mortality rates, public health authorities in England launched a campaign against the dummy: an infant welfare center in London even founded an Anti-Dummy League, which only accepted mothers whose babies had been pacifier-free for six months. Government health visitors, who paid house calls on new babies, told mothers in grave terms about the profound dangers posed by the dummy. It worked: in the English county of Hertfordshire, the number of babies using pacifiers was cut in half between 1911 and 1930. But a recent study, using data collected by the health visitors, shows that the babies who did use pacifiers were no more likely to be sick. Pacifiers were a phantom problem.

By the middle of the century, the pacifier was discussed in such different ways it is hard to believe people were talking about the same object. They were praised as a blessed godsend, the only thing that could stop the endless crying; damned as a source of pestilence; or grimly accepted as both—a necessary, noxious evil.

Benjamin Spock, the newly famous author of *The Common Sense Book of Baby and Child Care,* argued for a progressive attitude. In his trademark style, he talked parents down from their fears: the pacifier was rarely habit-forming, he stressed, and easier to take away than a thumb. And Spock had a keen understanding of the real problems of his readers—he told them what to tell their mothers: "If you feel that your baby needs a pacifier and are worried only about what the neighbors or relatives will say, tell the neighbors that this is a very modern practice (or tell them that this is your baby)."

Spock's attitude toward pacifiers was ahead of his time and it no doubt swayed some middle-class American households. But parental attitudes toward pacifiers, as with so many questions of child rearing, cleaved by class, and in general, the higher the class, the greater the horror. The fault lines are clearly outlined in *Infant Care in an Urban Community,* a remarkable book by the British psychologists John and Elizabeth Newson, who went around a British town in the late 1950s trying to find out how parents actually parented. The question "Has he ever had a dummy?" was met with "indignation among some middle-class mothers that the question should have been asked at all." On the other side of town, "a labourer's wife" who was asked the same question ended up apologizing because her child had never taken to the thing: "We tried everything—syrup, sugar, jam on it, everything, but he didn't want it." But many mothers were somewhere between these extremes, and in their responses, you can sense them straining to see through a miasma of conflicting information and impressions. And sometimes you can feel a mother's palpable guilt for giving in.

"I mean, when we first had her," a truck driver's wife told the Newsons:

> it was a fortnight of misery—cry, cry, cry—and I kept saying, I'm not going to give her a dummy—but in the end, of course, we had to do it, it was the only way we could get a bit of peace . . . One day the doctor came in; he says, Oh, she sucks a dummy does she? he said, You want to chuck that on to the back of the fire, he said, straight away, he said, she doesn't need that. Let her cry, he said, she'll get over it. Well she was poorly, she was full of cold and she'd got diarrhoea, it was her teeth the doctor said. She broke out all in this rash, and she was crying nearly all the time; and I thought, Well, I must stick it out, she's not going to have it today. But in the end I had to give it to her—it was the only way.

You end up wanting to chuck the doctor onto the back of the fire.

The class lines traced by the Newsons are still present in how people talk about pacifiers. Child rearing has always had these fault lines, of course. A century ago, the idea that the poor were criminally bad parents, whose toxic offspring were polluting society, was commonplace. The lower classes, it was said, needed to be taught how to raise children; they needed to be enlightened by their betters.

Few people say things like that anymore, at least not publicly, but you can hear their echo in much of the medical literature on pacifiers, which tends to dismiss any mothers who use them as ignorant and inadequate. The core assumption of the literature, summarized by an academic outside of it, is that pacifiers are used by "a young, uneducated, working-class mother who lacks both intelligence and educational qualifications, maybe a single mother, and is too lazy to interact with her child." Mothers who use pacifiers are seen as not really being *there*.

I had always thought pacifiers were shameful. A child with his thumb in his mouth—his warm skin, not some silicone—looks comforted. He looks at home with himself. A child with a pacifier looks plugged up. He's not pacified; he's *been* pacified.

Several weeks after Isaiah started sucking his pacifier, we went to a friend's house for dinner. Isaiah screamed almost the whole way there, but when we pulled up he'd just fallen asleep, the pacifier between his lips. He looked happy. He'd finally found a moment of peace. He was barely two months old.

"Okay," I said to Anya, "let's take the pacifier out." I said it casually, quickly, the way you'd tell a friend their fly was down. Of course you'd want to zip up your fly! Of course you'd want to take out the pacifier! Who walks around like that?

I got the sort of look that's rarely seen outside of divorce court. The pacifier stayed in.

Why did I feel so ashamed—or more precisely, why did I feel like I *should* feel ashamed? I was happy he'd fallen asleep with the pacifier; I just didn't want to take him out in public with it. In the United States, some two-thirds of infants use pacifiers at some point. Not only were we not alone: we were in the majority.

But I hadn't concocted this churning feeling, and I wasn't the only one who felt it. Pacifiers have an unsettled status: simultaneously popular and deplored. Given the problems they are accused of causing—language delays, early weaning, nipple confusion, dental deformities—such a schizoid response might make sense: you love them but you really shouldn't. But it doesn't make sense. Because when you scrutinize any of the accusations against pacifiers—the breast-feeding complications, even the mundane dental malformations—they collapse.

Everyone knows, for example, that pacifiers impede nursing. The Baby-Friendly Hospital Initiative of the WHO/UNICEF explicitly states that hospitals should "give no pacifiers or artificial nipples to breast-feeding infants." This prohibition seems intuitively correct: the flimsy, fake nipple must confuse infants and disturb the natural rhythms of nursing.

But the evidence for this claim is oddly underwhelming. In fact, the best studies on the question conclude that pacifiers, at least if given fifteen days after birth (and possibly even if given earlier), have no effect on the duration or success of breast-feeding. Many lactation consultants point out, correctly, that some studies have correlated pacifier use with early weaning. But pacifier use may be correlated with weaning without being responsible for it. The best randomized studies on pacifiers and nursing validate this: pacifier use is a *marker* of nursing problems, not a cause of it. In study after study, women who are motivated to breast-feed aren't affected by whether or not their infants use pacifiers. It's impossible to be definitive—better, bigger studies remain to be done—but judging from the best available evidence, pacifiers are simply not the bogeyman behind low rates of breast-feeding or early weaning. (In fact, a recent study from a hospital in Oregon found that when the hospital restricted the use of pacifiers, in accordance with the Baby-Friendly Initiative, breast-feeding rates went *down*.) Nipple confusion, for that matter, may be just a myth: there's almost no actual evidence for it.

But there's real reluctance to acknowledge that pacifiers might be benign. The current edition of the standard reference for lactation consultants, *Breast-feeding and Human Lactation*, states, flatly, "Pacifiers undermine exclusive breast-feeding for the first six months." Only studies that support this conclusion are cited; the many studies suggesting otherwise are ignored. You can sense the legacy of earlier pacifier panics in all this, the echo of the idea that pacifiers are—somehow, surely—*bad*. And then there's the fear, which permeates almost all advice about nursing, that if women take a small step away from breast-feeding—if they use a pacifier, or an occasional bottle—they'll never go back again. It's a slippery slope, and it always ends in the formula aisle. Of course, this logic only works as long as it scares the mother straight. For mothers who use an occasional bottle or pacifier anyway, the argument is counterproductive: they're basically being given a "scientific" reason to stop nursing.

Pacifiers are also blamed for delayed language development, which seems logical, too—how's he going to talk with *that thing* in his mouth?—but there's no evidence for this either. There is evidence that the lack of evidence hasn't stopped people from making the claim: a British speech therapist even admits she was *disappointed* her study's data showed no link between pacifiers and speech problems. And teeth? Pacifiers only screw up the palate if used past the age of five, well after the vast majority of children have stopped.

The hollowness at the center of the debate explains why, when the subject turns to pacifiers, a lot of sensible people stop making sense.

Amid all the warnings about pacifiers, it is startling to discover that the bulk of recent research on them is not about risks. It's about *benefits*. Pacifiers—somewhat mysteriously—reduce the risk of SIDS: the American Academy of Pediatrics now recommends pacifier use at night and during naps for infants up to a year old. (The controversial recommendation, issued in 2005, met with a firestorm of criticism from lactation consultants.) Pacifiers are highly effective pain relievers: they sharply reduce crying during painful procedures like circumcision. Since many pain-relief drugs can't be given to infants, pacifiers are often used as an anesthetic. They also seem to promote what scientists call better state regulation—what parents call an easier baby.

Pacifiers have a profound effect on premature infants, which is so unexpected it took a long time for doctors to figure it out. Why would a premature infant, whose nutritional needs are met intravenously, need to suck at all? But premature infants, given a pacifier, cling to it like a life preserver—some will suck at it determinedly for an entire hour. A series of studies has shown that preemies given pacifiers do better—they gain weight faster and leave the hospital as much as a week sooner—than those without. Pacifiers are now a standard component of preterm care.

A recent book on infant development, after weighing the pros and cons, arrives at this unconditional assessment: "Pacifiers provide comfort, promote physiological tranquility, and help in growth and development." If pacifiers are benign, or even beneficial, it is hard not to feel that what permeates the warnings about them is a fundamental distrust of parents: the fear that pacifiers will allow parents to detach themselves from their children—to substitute a cold, industrial object for warm skin and sweet whispering and a steady heartbeat. It's a modern version of Lindner's worst-case scenario.

La Leche League and the *What to Expect* books are even explicit about this fear: the pacifier, they warn, cannot substitute for a mother. This is the rare piece of parenting wisdom that manages to be both condescending and confusing. Condescending because it seems unlikely that parents who were considering using a pacifier—parents diligent enough to *look it up in a book*—were also considering abandoning their child altogether. Confusing because, well—*what?* How would a pacifier substitute for a mother— how exactly? Are there pacifiers on the market that cuddle and feed and rock and dote on a child? Is a mother nothing more than a nipple? In its sudden escalation of the stakes, the warning suggests there's something much deeper beneath the surface: the belief that the pacifier is the tipping point to bad parenting, if not outright child neglect. To give your child a pacifier is to temporarily separate yourself from him, and in that brief separation are the seeds of abandonment. In many contemporary books about babies, there's a fetishization of closeness for the sake of closeness. It's as if the whole complicated enterprise of parenting can be reduced to a single variable: proximity. The pacifier challenges that correlation: the pacifier represents distance.

Blessed, blessed distance.

I'd like to think that while Isaiah used a pacifier, we had *more* of ourselves to give him: screaming depletes parental love; it doesn't strengthen it. Of course, the current research on pacifiers might turn out to be flawed. Or maybe too many parents might rely too much on pacifiers. Or who knows what hypothetical scenario might come to pass. But until any of that actually happens, it'd be nice for parents to be told that their decision might not much matter. For too long, how babies suck has mattered way too much.

The long debate over pacifiers and thumb-sucking is yet another example of culture going to war against evolutionary biology. Think of how colostrum has been rejected or misunderstood. Refusing to allow babies to comfort themselves is nearly as senseless. Crying, especially in early infancy, the most vulnerable period in life, is exhausting: it saps an infant's energy resources at a time when those resources don't run deep. This is why, as the psychologist Elliott Blass notes, "multiple defenses have evolved to minimize it" and the most successful of those defenses "mimic those facets of the mother that are involved in nursing/suckling exchanges." It's not by accident that sucking is what stops crying.

The evolutionary logic of sucking is the central irony in a story that's crowded with ironies. Thumb-sucking isn't a neurotic, autoerotic, or nonsensical act. It isn't even strange. It's the baby trying to do what comes *naturally*. As the anthropologist Sarah Hrdy writes, "If a baby finds a rubber pacifier soothing it is because for at least fifty million years, primate infants so engaged could feel secure, because a baby sucking on a nipple is a baby likely to have a mother close at hand."

We suck on pacifiers because we're trying to find our way home.

The End of Orality

We took away Isaiah's pacifier at six months. The process was sudden and total, and although we felt bad for him, we managed to feel worse for ourselves, which is how we survived the screaming.

I should explain. His sleep had slowly degenerated, changing from the longer stretches of a baby with an expanding stomach to that of a baby with a deep need for his parents at every hour of the night. It wasn't because he missed us. It was because he missed his pacifier. When he lost it, he'd howl with betrayal. We'd soothe him and jiggle him and sing to him, which never worked, and then give him his pacifier, which always did. It wasn't like taking care of a baby. It was like putting quarters in a washing machine.

We needed more sleep. (Specifically, Anya needed more sleep.) There's a lot of understandable anxiety over letting babies cry it out—what's known, semi-accurately, as Ferberizing, which sounds reassuringly like an office carpet-cleaning service—but when your wife tells you that lack of sleep is dragging her into a depression, a few hours of screaming seems like a psychic bargain. And if we were going to ignore him, we had to ignore his pacifier.

Twenty-four hours in the life of a baby is a very, very long time. Within a day, he stopped asking for it. A day after that, it might as well have never existed. A couple of months later, we packed a pacifier for his first airplane trip, in case things got rocky. Things got rocky. But he didn't know what to do with the pacifier. He tried chewing on the handle, but it didn't occur to him that he was supposed to suck it. A few months before, such an oversight would have been inconceivable: it would have been the *only* thing that occurred to him.

He was developing other ways to explore his world: the arm, the hand, the fingers. Weaning happens at wildly different ages—the question of when a child is evolutionarily "meant" to be weaned is a very charged question—but sucking loses its primal importance early, around the time the rest of a baby's body catches up with his mouth. After the opening months of infancy—the time of Slinky necks and middle-distance stares and noses in mouths—there may be no instinct we ever experience that is as compelling, not to say tyrannical.

Isaiah's drive to suck—the drive that had defined his existence—had slackened. Better yet, it had been subsumed: the rest of his developing self had swallowed it up. He'd moved on from the first definable stage of his life—the stage when life is what happens between your lips—without us realizing it.

He couldn't even crawl yet and I was already trying to catch up.

Part Two

SMILE

What you doing?

Huh?

What you doing?

Look at all the people.

You want to look at all the people?

Huh?

You want to look at the sun?

Huh?

You don't want to look at the sun?

You want puppy dog to give you a kiss?

Give me a kiss.

See.

Oh, what was that?

Puppy dog give you a kiss?

Puppy dog give you a kiss?

Yeah.

Yeah.

Yeah.

You don't want to look nowhere else.

Huh?

No.

You just want to take a shit.

And then you'll be fine this afternoon, if you can just go.

Huh?

Yeah.

You'll just be fine.

Yeah.

You going to smile again?

Is Jennifer going to smile?

Is Jennifer going to smile again?

Huh?

No?

You're not going to smile?

You're not going to smile?

You're not going to smile?

Why?

[etc.]

—Mother talking to her six-week-old daughter,

as quoted in Kenneth Kaye's

The Mental and Social Life of Babies

6

The Birth of a Smile

Days after he was born, when we finally had time to talk it over, Anya said our newborn son had reminded her of a character in *Star Wars*—specifically, the gray-fleshed alien at the interstellar bar with the tube wrapped around his neck.

That's how Isaiah had come out, more or less. Without the doctors realizing it, his umbilical cord had somehow slipped into a noose, and although he was fine—the cord was cut immediately and his heart rate never dropped—for a moment he looked like a figment of George Lucas's imagination. He was fuzzy, half dismal gray and half blood-soaked red, and his eyes were open and rapt; they had an intensity that was almost terrifying. Seeing her son for the first time, Anya said, she didn't have the thoughts she'd expected to have. She didn't have the thoughts other people say you will have. What she thought was: *He's coming for someone and I hope it isn't me.*

And I hope it isn't me.

It's okay to say this sort of thing, I think, in the extended, elaborate framework of a *Star Wars* analogy—the joke's clear.

Outside of it, not so much. Even in the current era of deeply disil-
lusioned parenthood—when there are columns titled "Bad Parent" and features with headlines like "All Joy and No Fun: Why
Parents Hate Parenting"—ambivalence at the sight of your freshly
delivered newborn is the last taboo.

But should it be?

In her last trimester, Anya was diagnosed with gestational diabetes. It's a temporary condition—when the placenta goes, the diabetes goes with it—and it isn't passed on to the child. But there's
a small chance that the fetus can grow too big and unwieldy for
the birth canal—the shoulders bulk up like a football lineman's.
(Multiple doctors made this comparison. "Like a lineman's" is apparently some sort of official medical metaphor.)

Understandably, therefore, doctors are very nervous about
women with gestational diabetes going past their due date. Even
though Anya's gestational diabetes was well controlled, when she
was near forty weeks, her doctors pushed for induction. We had
reservations and we knew induction came with its own set of
problems, but there are only so many times you can hear someone say the words "increased chance of stillbirth" before you give
in. A couple of days after her due date, Anya was being IV'd full of
Pitocin, the commercial name for the hormone oxytocin, designed to kick-start the contractions of labor.

This is when things began to go awry: after a day on a Pitocin
drip, labor hadn't started. Days before this, Anya's doctors had already applied prostaglandin gel, a hormone-like substance meant
to soften her cervix. Now they used a different form of prostaglandin, a pill called Cytotec, which is placed at the opening of the
cervix. This is, somewhat bizarrely, an off-label (but common) use
for Cytotec: the FDA has only approved it for treatment of stomach ulcers. The Cytotec is one of many reasons why Anya would
later describe giving birth this way: "My entire experience of labor was of strangers coming in and sticking things up my cunt."

The miracle of childbirth!

If you're confused by all the hormones, and the hormone-like things that aren't actually hormones, and the nonapproved use of a powerful pill placed inside my wife and an inch from my child, that's the point: modern childbirth is a whirlwind, a maelstrom of pharmaceutical brand names and ominously beeping monitors. It plays as tragedy and farce at the same time: at some point, doctors burst into the room, ordered Anya onto her hands and knees, and clamped an oxygen mask on her face—the Pitocin was suddenly working with such strength it had depressed Isaiah's heart rate. They immediately counteracted the Pitocin with another drug, which worked too well: it stopped Anya's contractions. So they had to counteract the drug that counteracted the Pitocin by using—yes—yet more Pitocin.

Anya gave birth almost forty-eight hours after induction began.

I go into all this not because I think her labor was particularly unusual or lengthy or gruesome—as far as I can tell, it wasn't. And not because I think her doctors screwed up—as far as I can tell, they didn't. But the story of labor is often reduced to pain and reward: the agony and the ecstasy. This is too easy. Childbirth doesn't have a clear narrative. It's all chaos and uncertainty and confusion; it is, ironically, often dehumanizing. From this, love is supposed to instantly bloom. It's hardly surprising that sometimes it doesn't.

In the mid-1970s, the young English academic Ann Oakley interviewed dozens of women about to have their first child at a London hospital. Oakley spoke to the women several times, before and after they gave birth, and she was so struck by what she heard that she published a very unacademic book, *Becoming a Mother*, composed almost exclusively of the women's own words. In the category of books about pregnancy and parenthood, where emotions are usually prescribed for the reader, it remains a rare, raw book—a book that admits things can be complicated. Here are several women on how they felt upon first seeing their child:

Anne Bloomfield, 23, barmaid: "It was really weird. I had no reaction, I didn't feel anything. I didn't care about myself, I didn't even care about the baby. They put the baby in my arms and I just looked at her. It was really weird."

Alison Mountjoy, 27, fashion designer: "It was very strange. I mean yes: it was *my* baby, and I loved her, but I think I was just so shattered by then that whatever I was feeling I couldn't feel much of. I mean I was totally aware that she was my baby and I loved her and I wanted to hold her but I felt so sick I couldn't react to *anything* by then."

Pauline Diggory, 25, market researcher: "They gave her to me: I felt bewildered. She was looking at me—I was looking at her. She looked very bewildered herself really."

Sue Johnson, 27, photographer: "I really felt she was a stranger. It took me really a long time to get to love her really. It's such a funny thing, that suddenly you've got this person that you've got to get to know . . ."

Sasha Morris, 26, air hostess: "I was very amazed at my own reaction when she was born: I was completely *numbed:* I thought I'd be delighted. I think a lot of people won't admit to their feelings. They say they're absolutely delighted, but I'm sure half of them aren't. It's quite normal, isn't it?"

Weird, strange, bewildered. It's the vocabulary of women trying to align their expectations with what they actually felt.

"After such hard labor," Oakley asked, "is anyone in a fit state to fall in love?" Isn't it possible that love is the emotion a new parent is sometimes *least* prepared to feel?

To be the father in the delivery room is to be really, really lucky. But watching the person you love most in the world experience excruciating, seemingly unending pain—while being unable to do anything about it—has its very small price. When Isaiah was born, I was, if not numb, like Sasha Morris, then muted. I'd been overwhelmed by emotion—there was so much of it, I'd stopped letting most of it through. What got through, strongly, was relief: it was over. He was here.

Love sometimes comes in swells and sometimes in small, Seurat-like dots—you don't know it's there until you step back from it. With Isaiah, I expected the former and got the latter. I loved Isaiah well before he ever smiled. But the first time I felt the sharp upswelling of love I'd expected at his birth—a rogue wave of emotion—was when he smiled.

Here's how I described it at the time. It's impossible to capture the blotchy purpleness of early parenthood even a few years afterward. Sooner or later, you've got to go back to the diary. We'd set up a blog for the grandparents and other interested parties, and on January 17, 2009, I posted this:

Last Wednesday was I's six-week birthday. It seems like just yesterday he was a baby.

Six weeks is when I's supposed to start smiling like he means it. Up until now, he's been smiling, but he hasn't known it: according to The Books, smiles before six weeks correspond strictly to changes in sleep states. (The Books can be merciless.) When I drifts off, a half-smile will sometimes break over his face: the effect's a little like he's enjoying a private joke.

It's another example of how evolution seems to have inverted the proper order of things: the screaming would be easier to take if he threw us a smile now and then. But over the last week, he's been trying on new facial expressions, and if he's in the right mood and you're being the right sort of silly, his mouth will open up into a toothless half-moon. For a second, it pulls the tides along with it. And then it dissolves.

You've only read this far because you're expecting a photo, . of course. Sorry. Soon.

The next post was just a photo with a heading: "And There It Is." We immediately printed fifty copies of that photo. It took a lot of self-restraint not to hand them out on street corners or stick them under windshield wipers.

The Person in Your Child

Looking back, it seems like a comically exaggerated reaction to a single smile. But this is what's most strange: it was not that strange.

"Parents observing for the first time their infant smiling while gazing at them or in response to their own smiles," the psychologist Philippe Rochat notes, "discover a person in their child: a person among other persons. It is invariably experienced by parents as a memorable first greeting of their child." Rochat's phrasing—*a memorable first greeting*—skips quickly past the first six or eight weeks of life: the first exchange of smiles allows parents to meet their child as if for the first time. It's the *real* first greeting. The moment when a baby announces his presence with a smile is, as the psychiatrist Daniel Stern says, "almost as clear a boundary as birth itself."

The Anbarra, an Australian aboriginal people, treat this moment *as* a birth. After a baby is born, the Anbarra continue to call him or her the same word they use for the unborn, *yukiyuko*. For the Anbarra, a person's true entrance into the world takes place not at birth but weeks or months later. It's not being out of the womb that matters; it's smiling. When a *yukiyuko* begins to smile, it stops being a *yukiyuko*. It becomes a boy or a girl, an *andalipa* or a *djindalipa*—a child.

For the Anbarra, smiling is what marks the arrival of *humanness*. This is pretty much precisely how I'd felt about Isaiah. Suddenly, the creature before me had turned into a person.

Why do we find so much meaning in a half-moon of a mouth? The smile is a shifty subject: once you start trying to figure out what's behind it, no assumption seems safe to stand on. We tend to think of smiles as pure reflections of happiness; almost all cultures make the same assumption. But it falls apart after a few seconds of reconsideration. After all, we also smile out of discomfort or awkwardness; we smile to appease someone more intimidating or powerful. We smile when politely trying to wiggle out of a boring conversation.

Even if we smile for other reasons, though, we still break into a smile when we're happy. Or do we? When people go bowling, a famous study found, they don't smile after a strike. They smile when they see other people who saw their strike. People who bowl alone barely smile at all. A similar study was done of the award ceremonies at the 2004 Athens Olympics. It concluded that gold medalists, people who were in theory as happy as they possibly could be—in theory way happier than people at a bowling alley—did not smile. Or more precisely, they only smiled when they interacted with other people. The social interaction, not the fact of the gold medal, is what sparked a smile.

This is true in more mundane circumstances, too. People smile much more when watching a video with a friend than when alone. Even *implied* social interaction produces more smiles: if people are alone but told that a friend in the next room is watching the same video, they smile as much as they did when their friend was in the same room. "There is no point in being able to smile," the psychologist Vasudevi Reddy writes, "unless there is someone out there to feel its impact."

Smiles aren't simple things. To quote Melville, who saw this plainly: "A smile is the chosen vehicle for all ambiguities." It's not a reflection of inner happiness, and inner happiness isn't manifested in a smile. We're dimly aware of this disjunction, since we unconsciously act according to it—we smile out of something less than inner happiness. But when we attribute meaning to the smiles of others, we elide it: we want to believe smiles are meant in earnest. And there's no time we want it more than when we look at a smiling baby.

But to ask what a baby means by a smile is a stubbornly complicated question. Does the smile reflect social engagement? Or stirrings of emotion? Or some measure of happiness—and if it did, what would a baby even mean by happiness? If you ask Daniel Messinger, a professor at the University of Miami who studies infant smiling and probably has spent more time looking at smiling babies than anyone who's not a children's entertainer—if you ask him what a baby feels when she smiles, he'll pause and then say, "That's a good question and that's a hard question." And how could it not be? We often don't even know what adults mean by a smile, and they have the unfair advantage of being able to tell us.

"On the one hand, it seems clear that smiles have some irreducible emotional meaning," Messinger continues. "On the other hand, we know they can be used for all sorts of rather complex things. On the one hand, we all know a smile when we see one. On the other hand, there can be a lot of different kinds of smiles." This is how complicated smiling is: it has *four hands*.

When it comes to smiling, though, there are far fewer vivid cross-cultural accounts of weird child-rearing practices—there are no stories about societies that encourage frowns and forbid grins. That's not to say everyone has had the same reaction to a baby's smile. Some cultures value smiling less—Nso mothers in Cameroon, for example, don't spend much time encouraging the expression of positive emotion; they spend a lot of time discouraging the expression of *negative* emotion. They care most about calmness, not delight. There are no historical studies on varying attitudes toward infantile happiness. Did the Puritans, who thought deeply about babies but also believed they were damned, see a baby's smile in a more ambivalent way than we do? We don't know. We only know that, in the historical record, when smiles have flashed across the faces of babies, they have brought delight.

Unlike seemingly every other aspect of infancy, in which unimaginable variations are the norm, smiling is by and large a story of dully reassuring sameness.

Inside the laboratories, though, the story is very different. If sucking is an exhausted, depleted subject of study—an act that was supposed to be at the heart of infancy and ended up somewhere closer to the appendix—smiling is the opposite: the focus of a lot of rewarding developmental research taking place right now. Writ small, as the study of when infants smile and how those smiles change, it occupies a small but absorbing body of research. Writ large, as the study of interpersonal communication in its earliest stages, it underlies foundational questions about how early experience shapes development. It stretches even further than that, though. It wanders into the sort of questions that bring on vertigo. What is it that makes us different? What is it that makes us essentially *human*?

These are the questions of well-rested scientists, however. For the exhausted parent, a baby's smile has such an impact because it seems to emerge out of nowhere. Time with a newborn is distended past all recognition: a day feels like a week; a week feels like a year. By the time a baby hits six or eight weeks—the age when smiles appear on the horizon—it can seem like you have been grinning madly at this creature for fully half your adult life. Has he even been paying any attention at all?

Making Faces

When babies are born, they don't look at the heart rate monitor, or the view out the window, or the clock. They look at faces. This might seem unimpressive: when you enter a room full of people, you look at faces, too. But you've had years of practice. Even though a newborn has never seen a face before, she's still attracted to faces—she somehow knows that faces are special.

Faces matter. They are a baby's biggest source of information about the world. And at birth babies have a tentative but very real preference for the feature of their environment that will turn out to be the most important.

This interest in faces isn't learned. To confirm this, an experiment tested babies who'd had no time to learn—their median age was a stunning nine minutes. The infants were more attentive to a pattern that resembled a face than they were to scrambled versions of that pattern or to a blank sheet. Only nine minutes removed from the womb, babies will turn their eyes and heads to follow something that looks like a face.

It's a measure of how important faces are for infants that, despite absurdly limited vision, they are still able to distinguish among them. Newborns only a few hours old prefer pictures of their mother over pictures of other women. They also prefer attractive faces: babies look longer at people who are better-looking (as rated by adults, obviously). This finding is depressing but not as strange as it might seem. It turns out that when the features of many, many faces are combined into a single face, the result—the average face—is always very attractive. If newborns have a prototypical face in mind, in other words, that face is attractive. What we see as handsome, a newborn sees as average. That's why babies like beautiful people: they are what they *expect* to see.

Babies turn toward beautiful people but they also, as the months go by, begin to turn toward people who are most like the people they know, regardless of how good-looking they are. This effect is called perceptual narrowing—the disheartening fact that what we can perceive is limited by what's around us. If babies don't hear certain sounds, they stop being able to hear those sounds: sometime toward the end of the first year, infants lose the ability to discriminate among phonetic variations in a foreign language.

The same fine-tuning happens with faces. A few months after birth, babies raised mostly by women prefer to look at female faces and babies raised mostly by men prefer male faces. Because most infants spend more time looking at female faces—because there are more women than men taking care of babies—they comprehend them better: babies, at least those raised primarily by women, tend to see female faces as individuals and male faces as a category. (Women have identities; men are just *men*.) Babies are, as a few psychologists have put it—in a phrase that's more Greek system than ivory tower—"female experts."

At this point babies are still perceptually "broad" enough that they can discriminate among the faces of monkeys as well as they can people. But as they approach their first birthday, they lose this talent: they reliably distinguish only human faces. They have stopped seeing what's not part of their world. It's a sad moment—it may be strictly neurological (the pruning of the perceptual system) but it feels philosophical. From this point on, like the rest of us, babies only see what they already know.

A newborn doesn't just know that faces are important. There's ample evidence that a newborn can *imitate* faces. But how?

This question quickly drags you into deeper waters. I discovered this recently, when I tried to explain to some friends why the phenomenon is so weird: "So pretty obviously, a newborn baby has never seen herself. She's never seen a picture. She's never seen a mirror. But newborns can somehow imitate facial expressions. You stick your tongue out, they stick their tongue out. Which means a lot. It means not just that they know they have a tongue but that they know their tongue is like your tongue. And that they know how to *move* the thing they have that is like your tongue. Which is weird because although they can feel their tongue move, they can't feel *your* tongue move. Which they shouldn't even know they also have. And here's the weirdest thing of all: newborns mostly imitate with their faces, which is the part of their body they *can't see*."

I stumbled along like that for too long. I must have sounded stoned. The weirdness of neonatal imitation, if you let yourself get lost in it, leads inexorably to: *Whoa. Dude.*

And for a long time, the claim of neonatal imitation would only have made sense if you were in a mind-altered state. Any sober psychologist would have dismissed it outright. Infants were considered capable of imitation at a year, not immediately after birth. It was an intellectual achievement, hard won after many months of development, not a response within the repertoire of a neonate.

That idea was upended in 1977, when a young psychologist named Andrew Meltzoff, working off an observation of neonatal imitation made a few years earlier, decided to test the claim in controlled conditions. In a paper that's been cited nearly two thousand times, a discipline-altering number, he reported that babies as young as twelve days old will readily imitate an experimenter who is willing to make a fool of himself—sticking out his tongue or craning his mouth wide open.

Twelve days after birth was early enough to be shocking, and many people were shocked, but it wasn't early enough to prove anything—theoretically at least, a lot can be learned in a couple of weeks. To eliminate any doubt about what's innate and what's learned, you need newborns. And you need them *new*—the problem with newborns is that they get old fast. Meltzoff arranged for expectant parents to call him when their children were about to be born—"Honey, start the car and call the developmental psychologist!"—and tested the newborns in a small laboratory he'd set up next to the birthing room. After a year, he'd examined eighty infants and established that neonates have the innate capacity to imitate.

Neonatal imitation is simple: your tongue wags, my tongue wags. It's only impressive because of the neonatal part. Within a few months, though, infants imitate in far more sophisticated ways. A six-week-old who observes a researcher making a single facial gesture—opening his mouth, say—will reproduce that same gesture a day later, without prompting, when she meets the researcher again. Babies at this age remember what expression goes with whom: if their mother makes one expression and a stranger makes another, babies track who made what expression and reproduce each for the appropriate person. They seem to sense that the people around them are distinct and not fungible, human-shaped objects.

Precisely how babies can imitate remains a mystery, but a little over a decade ago, in a famously serendipitous experiment, a group of Italian neuroscientists excavated a neural foundation for the capability. The scientists were studying macaque monkeys—specifically, they were studying a small section of the premotor cortex, the part of the brain more or less responsible for action—when they noticed something seemingly inexplicable: the cells in a monkey's brain that activated when a monkey moved were also activated when a monkey *saw* the scientists move. When a researcher reached for an object, the monkey's brain reacted as if *the monkey* were the one reaching for the object. The brain—that's to say, the cells in this small part of the premotor cortex—behaved as if acting were the same process as perceiving: monkey see, monkey do.

These cells were named mirror neurons—since they mirror actions—and they provide a clue to the mechanisms of neonatal imitation. But mirror neurons may have a deeper philosophical significance, too: they let us, neurologically, at least, cross over the chasm that exists between the self and the other. They allow us to understand the actions of other people *as if we were acting ourselves.* (Luckily, there's also a mechanism that inhibits us from actually acting ourselves. Without this check, we'd be stuck in a cycle of endless imitation.) The existence of mirror neurons suggests that there's a deep neural basis for sympathy—for seeing oneself in someone else.

"The child, even the newborn," Meltzoff writes, "can watch the movements of other people and immediately recognize that

'those acts are like these acts' or 'that looks the way this feels.'"
This is a seminal moment of recognition: it is how we begin to
connect ourselves to others. As Meltzoff sees it, infants take this
initial, shaky sense that other people are *like me* and use it to discover themselves.

A baby's smile seems to appear out of nowhere. But that
"seems" is an illusion. Newborns—not just babies but womb-wet
newborns—have been paying attention, and trying to get attention, all along. Like the aliens they appear to be, they have been
trying to make contact.

7

"A Mother Evidently Perceives Her Baby to Be a Person Like Herself"

For years, the entire paradigm of infant development was built on the idea that babies weren't aware of people: it was what psychologists believed, it was what doctors were taught, it was what parents were told. The answer to the question "Is your baby paying attention to you?" was: No. It was bold, all-caps, 72-point No.

This is the potted-plant paradigm again—the long-standing assumption that babies weren't capable of much of anything. Jean Piaget considered the idea of an infant interacting—the idea that a baby would be aware of and attentive to people—to be preposterous. For Piaget, babies were fundamentally egocentric, incapable of taking anyone else's perspective: under his theory, it took at least eighteen months for an infant to see beyond her own shadow. For much of the twentieth century, as Andrew Meltzoff has written, "scientists often assumed greater similarities between college students and rats than between college students and infants."

A decade before Meltzoff made silly faces at newborns, the potted plant paradigm had started to crack. A few researchers, working with infants a bit older—around a couple months of age—had demonstrated that babies were not in fact hermits.

These researchers, who were about to revolutionize the field of developmental psychology, tended not to be psychologists. They were outsiders. And perhaps the most outside of the lot was an accidental radical with the storybook-like name of Colwyn Trevarthen.

Trevarthen had been an outsider from birth. Born in New Zealand, on the edge of the British Empire, he grew up skeptical: "I tell people when I was a young person I felt that Europe was a myth, and when I came I discovered I was right."

He had the childhood of a primitive anarchist, he says, agreeably. "I lived like an Amazonian Indian in the forest. I was an expert in going around in forests that are really something like a jungle." He was trained in zoology and as a field biologist, and he studied ecology, the science of how the parts of a system respond to each together—a perspective that would guide his later work on mother-infant interaction.

Trevarthen would go on to be a much-honored professor of psychology, but he never had any intention of being a psychologist. Even at the time he began work on babies—in the late 1960s, at the Center for Cognitive Studies at Harvard—he'd never trained as a psychologist: he had left New Zealand to do his graduate work at Caltech in neuroscience. Intellectually, he was about as distant from mainstream psychology as he could get—nearly as far afield as he'd been as a child in New Zealand. "I didn't even *know* about Piaget until I went to Harvard," he says. "I'd never heard about Piaget before I was already unable to accept his explanations."

Many psychologists at the time were interested in how infants perceived people. Trevarthen was interested in how infants *interacted* with people—how babies and mothers played off each other. His research technique for studying early interaction was barely a technique at all: sit a mother and her baby in facing chairs, ask the mother to talk to her child, film what happens next. Trevarthen was trained in natural history: he wanted to observe the mother and child acting naturally. "I wasn't interested in what they could do. I was interested in what they *wanted* to do."

What Trevarthen discovered—after comparing weeks of footage of how babies responded to their mothers and to a suspended toy—was that infants know people are fundamentally different from objects. Babies treat objects, he wrote, as "potentially graspable, chewable, kickable, step-on-able or otherwise usable." But they do not communicate with them *expressively*. In other words, babies *do* things with objects. They save their expressiveness for people. And they seem to expect people to respond.

When a baby can do nothing else, long before he can reach out and bat the mobile hanging right above his head, he can zero in on the people around him. He can develop, in Trevarthen's words, "a deep affectional tie" for them. At only a couple of months of age, a baby not only recognizes his mother but "invites her to share a dance of expressions and excitements."

Trevarthen called this ability "innate intersubjectivity." It's a clumsy but simple enough term: it represents the idea that infants *want* to dance this dance—that they are driven to share emotions and find meaning through interaction. It's through emotional exchanges, through these dances of expressions and excitements, that babies begin to sense the subjectivity of other people—their essential humanness.

Trevarthen was among the very few people who were not surprised by what he'd found. He'd already watched his newborn son with the eyes of a biologist: "I saw the self-awareness expressed in the small, delicately coordinated movements of his eyes, face and hands, and their readiness to meet in smiling moments with the expressions of his mother," he wrote many years later. "Babies, I felt, act like people and seek to communicate. Their impulses are sociable before they are practical or rational."

Just as mainstream psychologists were conditioned by their studies to not see mother-infant interaction, Trevarthen was trained to recognize it. "I was a very Darwinian biologist and it's absurd for a biologist or ethnologist to think that babies aren't social. Humans are hyper social, *extremely* social. They must have deep adaptations for that."

The idea of babies as social beings was not new. Darwin himself had looked at babies in the same way. After his first son was born, Darwin made detailed observations of him and repeatedly noted his awareness of the world: "An infant understands to a certain extent, and as I believe at a very early period, the meaning or feelings of those who tend him, by the expression of their features." At the turn of the century, the British philosopher G. F. Stout concluded that "intersubjective intercourse" must exist early on in infant development. But these old ideas had been buried by the potted plant paradigm.

The old ideas were now aided by new technology: Trevarthen used mirrors to record the face of the mother and the infant together on each frame, enabling him to analyze the tape in minute detail. Researchers like him could now decipher exactly how infants coordinated their responses with other people—a flurry of activity that was too fast to reliably transcribe as it happened. They could now see what was invisible in real time. The intersubjective theory was built not on experiments but on these observations. It was a social science, almost an anthropology of infancy.

In fact, Mary Catherine Bateson, the daughter of the great anthropologists Margaret Mead and Gregory Bateson, analyzed a film taken of a mother and her nine-week-old infant. This was the very opposite of experimental psychology—it was a work of almost ethnographic description. Like Trevarthen, Bateson's outsider perspective allowed her to see what others had overlooked: that the mother and infant were engaged in what she termed "proto-conversation." "These interactions were characterized by a sort of delighted, ritualized courtesy," she wrote. It's a lovely description that is immediately recognizable: it captures the direct, lively way in which even very young infants, when in the right mood, look their partner in the eye and fall into a conversation. They seem to know how a dialogue works: you go, then I'll go, then you again. Instead of following like-with-like, babies respond in the same emotional register: they follow a smile with a coo, or a coo with smile. Emotion is content.

Of course, we all assume a lot when it comes to babies—we have to fill in the silences. To a lot of people at the time, it seemed far more likely that what was happening was not proto-anything but *pseudo*: it wasn't real; it just looked real. The back-and-forth between a mother and her baby might appear to be a conversation but just be a set of disconnected, random responses. Early interaction, according to this way of thinking, was an illusion.

This is a scientific problem any parent can appreciate: what gets you through the early days of parenting is projection. How do you separate what you think is happening—what you swear is happening—from what's actually going on?

In part to answer this question, researchers devised the still-face paradigm. You can try this at home, although you might regret it: in the midst of a lively back-and-forth with a baby, stop responding, hold your expression still, don't look away. The effect is instantaneous: even a two-month-old baby will immediately sober up. She will tentatively smile, then look away, then look back, now wary but still trying to make contact. When she finally gives up, she will turn away, despondent. She will withdraw from the conversation.

The still-face scenario has a problem, though: it is too strange. When parents interact with their babies, and especially when their babies seem to be responding, their faces light up. They don't stay still. In the still-face scenario, a baby might freak out not because the conversation was inexplicably interrupted. He might freak out because the still face is so novel.

A brilliant experiment, designed by Lynne Murray, a student of Trevarthen's, controlled for this possibility. Murray had mothers interact with their two-month-old infants via video monitors—

the baby saw the mother's face on his screen, the mother saw the baby's face on her screen. After a few minutes—after the mother and infant had established what seemed like a dialogue—Murray cut the live transmission and replayed the earlier footage. The babies noticed immediately: the same behavior that had made them happy now left them puzzled. They smiled less; they looked away and then tried to reinitiate contact; they knew something was off.

These interactions weren't pseudo-anything. Edward Tronick, the psychologist who developed the still-face experiment, described the back-and-forth beautifully: "Never is one partner *causing* the other to do something. One musician does not cause the other to play the next note. In the same manner neither the mother nor the infant causes the other to greet or to attend. They are mutually engaged in an activity." Evolutionarily, this mutual engagement may be a solution to a stubborn problem—the fact that the newborn, born before his time, can't do much of anything on his own. He needs people close by who *can* do things. So he traps them in conversation. Through these conversations, babies learn about how adults—these crucial beings who can do things—work. These dialogues allow babies, in short, "to monitor and predict more accurately the behavior of those they depend on."

This new version of the infant was the sweet revenge of mothers past. Mothers had been told for decades that what they'd thought was a genuine dialogue with their babies was nothing at all—it was a naïve figment of their overactive maternal imagination. When psychologists overturned this idea, they did so by acting a lot like mothers: they paid attention to the mundane intimacies, the emotional conversations, of a mother and child; they described what they saw, plainly; and they did so in emotional terms, even though everyone "knew" those terms were inappropriate for infants. The mothers hadn't read their Piaget. They didn't know any better than to be right. Ultimately, the psychologists followed their lead. "A mother evidently perceives her baby to be a person like herself," Trevarthen wrote at the time. "Mothers interpret baby behaviour as not only intended to be communicative, but as verbal and meaningful."

Even the experts turned on the experts. "Until recently most scientifically minded people in our culture considered infants incapable of communication since they don't talk," the psychiatrist Margaret Bullowa wrote in *Before Speech,* an influential book on early communication. "Lately the infant has been discovered to be human and so more able to communicate with other humans than was formerly thought. Why should it need mentioning, much less discovering, that human young are human too? Simply because the 'experts' seem to have forgotten it for a time while they were busy being scientific."

Today, almost a half century later, the experts are still at war with the experts. There's no single version of almost anything in infancy. The field of developmental psychology has been split by numerous schisms, and each sect has their own theory-of-everything, each explanation privileging different factors.

Even when there is a consensus on what babies can do, there's no consensus on what can be concluded from it. Some psychologists believe that babies can communicate shortly after birth; others push it back to around six weeks of age. Still others argue that infants can't truly know what it means to communicate until much later, toward the end of the first year, when they are able to infer that other people have minds. Ultimately, the argument here is over what sort of thing *minds* are. Are minds opaque and hard-to-handle objects that have to be painstakingly theorized by babies? Or are they readily understood shortly after birth, no more complicated to grasp than the reality of hands or eyes? Do babies simply observe or do they have to interpret?

The case for infants having *some* sort of social sense is so strong that it cuts across all camps. Interpersonal consciousness was long thought to be the end point of infant development—its culmination. The revolution in infancy research has reversed this timeline.

It is now thought that this social awareness—the ability to sense that other people are people like me—is the starting point for development. Social cognition isn't the final product. It's how you make the final product.

It is hard to exaggerate the magnitude of this change. For a proper analogy, you have to concoct absurd scenarios—say, that art historians had concluded we'd been looking at the *Mona Lisa* wrong: it had been hanging upside down. The *Mona Lisa*'s smile belonged on the top and it wasn't a smile at all—it was a frown. That's roughly what happened to the conventional narrative of infant development: once it was flipped on its head, all the features looked different.

This is way more than an academic debate. How we see babies has always affected how we treat them. "It was not so long ago, after all," the psychologist Vasudevi Reddy observes, "that folk wisdom had it that human infants were born unable to see or hear, and that they were capable of little more than 'mewling and puking in the nurse's arms,' as Shakespeare put it. It was not so long ago either that medical science asserted (without parents being able to challenge it) that neonates cannot feel pain and thus justified a variety of intrusive practices like surgery without anesthesia." When we underestimate infants, even today, Reddy argues, we risk making the same mistakes. "If we think that infants have little in the way of thoughts and feelings and perceptions, then we will do less to look for or respond to what others might see as infants' thoughts or feelings."

Our expectations guide our actions. That's why the argument over early interaction mattered: the academic consensus filtered down and shaped what parents everywhere saw in the crib. An abstruse, awkwardly worded debate about whether babies are egocentric or innately intersubjective affects how people who aren't scientists react to a baby—even to a baby's first smile. It isn't just the baby who's deprived here. It's the parents, too.

Humans are attracted to smiles—we gravitate to the sight of a baby grinning. But that's not enough. We're also at the mercy of peer review.

What Academics Talk About When They Talk About Babies

Babies used to be interesting because they had parents.

For a long time, babies were an object of study mostly for the sake of the grown-ups. People wanted to know how to raise babies, or at least how to cope with them, or stop them from doing things, or get them to do other things. Much of the point of learning about infancy was to control and change the infants, partly for the benefit of their parents, partly for the supposed benefit of society as a whole. There are many exceptions, of course, but for many years research into infancy was ultimately about something other than the infants themselves.

It's not that people weren't curious. When child psychology was founded in the late 1800s, the adults had lots of questions— they wrote very detailed ethnographic reports of the life and times of the infant. But that research mostly revealed what the infant did, rather than how the infant *worked*.

This is no longer the case. In the last half century, research into how babies work has exploded. The so-called revolution in infancy was sparked when scientists figured out how to ask questions that babies could answer, if only inadvertently. They designed experiments around the key perceptual capacities that infants have. (They can look toward things! They can look away from things! They can suck!) And once the scientists started getting answers, they realized how many things they wanted to ask. They haven't stopped with the questions since.

Scientific interest in infants has never been more intense than the present, which is partly why there are such sharp schisms among the scientists. The theories proposed by Trevarthen, for example, are disparaged by the cognitive theorists, whose theories are in turn disparaged by people like Trevarthen. The story I tell about smiling is cobbled together from various theories, and most everyone will think I have misunderstood something critical, which I promise I have. But most everyone who studies infancy thinks that most everyone else has misunderstood something critical.

These academic arguments are so loud they can obscure the fact that, in the end, almost everyone who studies infants thinks roughly the same thing: these babies are *marvelous beings*. The various arguments about how babies work are basically arguments over the exact way in which babies are amazing.

If that sounds like sly sarcasm, it isn't. This perspective is genuinely new. Historically, babies haven't really been regarded as amazing—lovable, exasperating, mysterious, impossible, yes, but not *amazing*. The revolution in infancy research has changed how scientists, and in turn how parents, see babies.

At its core, the story of sucking was about how adults once believed infants had to be restrained and controlled—how their natural orientation toward their environment (to stick it in their mouth and suck on it) needed to be thwarted. It's the story of babies past. The story of smiling is about how adults believe infants are astonishingly capable and creative—how their natural orientation toward their environment is multifaceted and purposeful in ways we are only beginning to understand. It's the story of babies present (and possibly future). The stance a science takes toward its subject says a lot about the culture as a whole and everyone in it. In this case, science says, very simply: babies are worthy of wonder.

The Smiles in the Trees

I wasn't sure if Isaiah liked smiling, or if he liked how we responded to him smiling, or if those things were actually the same thing. But once he started, he couldn't stop: the experience of smiling was addictive.

This isn't a given. Babies who are blind begin smiling at the same age as other children—they don't need to have seen a smile to produce a smile. But their smiles diminish over time, declining in size and number. What they are missing are the smiles their smiles bring into being.

To survive, a smile requires other smiles. Detailed studies of mother-infant interaction show that these smiles—the ones in response to a baby's smile—mostly appear *after* a baby starts smiling. This isn't how we imagine the earliest mother-infant interactions—we think of parents as modeling expressions for their babies, who then adopt those expressions. We think of the baby as getting the smile from the parent. But until a baby smiles, an infant's blank look is often mirrored by his mother's blank look. The mother doesn't light up the baby. It is the baby, when his face comes alive, who lights up the mother.

Strategically, Isaiah's smiling was very clever. A few weeks before, he'd paid attention to us periodically, with this look of half-recognition, as if he knew us from somewhere but couldn't quite place where—*Was it the hospital?*—but he preferred the wall or a spot just over our shoulders. You always had the feeling there was someone more interesting standing behind you, as if he were searching for a better prospective parent. It's hard to fully engage with someone who appears to be avoiding you. But after a few simple smiles, we spent our days with stupid grins frozen on our faces: with a few easy, cost-efficient facial gestures, he'd snagged us.

What I remember from those weeks: long stretches of nose-to-nose silliness; imitating his ridiculous, nonsensical expressions; teasing him until his mouth parted in a smile. These early face-to-face exchanges, Philippe Rochat argues, provide babies with "a running commentary about how they feel and what they experience, whether joy, pain, sadness, or excitement."

Before these nose-to-nose exchanges, I'd never paid much attention to the complexities of the face. Most of us don't; we coast on language. The psychologist Silvan Tomkins, who pioneered the modern study of emotion, was legendary for his ability to read faces—to tell what people *meant* to say, not what they were saying. When Tomkins began researching how the face functions, he watched television with the sound off. He didn't want the words to distract him from what the faces were telling him.

If we're not Silvan Tomkins, we don't do this sort of thing. Babies *force* us to do this sort of thing. There are never any words to distract us from their faces: we all watch babies with the sound off. Their faces say a tremendous amount: they are flush with feeling and with meaning. To a smitten parent, a word like "wisdom" isn't too big for a baby's smile.

On the one hand, that's awfully precocious for a six-week-old. On the other hand, it is six weeks and then some: these facial expressions have been refined for millions of years—long before there was even humanity.

Among primates, those who live in more social environments have more complex sets of facial muscles. A more complex set of muscles means a more flexible face; a more flexible face means a more expressive face.

Chimpanzees, who live in a society nearly as social and fluid as our own, have a facial musculature that hardly differs from humans'. Like humans, they are good at telling individuals apart, even as infants, and like humans, they live in a social world that's in flux: they may spend long stretches of time away from people they know well. Such a society can't function if its members aren't able to detect even very subtle differences among faces.

To negotiate this highly social, highly mobile world, chimpanzees use facial expressions in a way not unlike our own. Among their everyday expressions is the silent bared-teeth display, the look they produce when they retract their lips from their teeth and gums, lift up the corners of the mouth, and generally assume the exaggerated appearance of someone with a shit-eating grin. It looks resolutely fake. It's a pawn-shop smile. But it may be the real thing—the original smile.

The bared-teeth display appears aggressive, but its purpose is conciliatory. After a chimpanzee flashes his teeth at another chimp, each becomes much more amiable. It's as if the bared-teeth display reduces the tension of an interaction: it makes it easier to get along. A flash of bared teeth represents, in the language of primate researchers, benign intent. Instead of an attack, you get grooming.

Baring teeth in the chimpanzee world, in other words, sounds a lot like smiling in ours. The bared-teeth display operates on a continuum, though: in more hierarchical primate societies, it is a gesture of appeasement; in more egalitarian societies, it is a gesture of cooperation. (Until recently, bared teeth were considered a "fear grimace," a conclusion researchers made mostly from observations of rhesus monkeys. Rhesus monkeys, however, have an unusually hierarchical social structure and their submissive use of bared teeth is very different from that of many other primates.) The more communal world of our chimpanzee ancestors stressed the cooperative side, and over the epic timescale of human evolution, the original sense of cooperation crept closer to pleasure.

This connection was once radical: when creationists attacked *On the Origin of Species,* they used our facial expressions as prime evidence that humans had not descended from anyone, especially apes. Our superbly complex set of facial muscles allowed us to express "higher" emotions than apes; it was proof that God had raised us above the lower emotional lives of mere monkeys.

Darwin's answer to this argument was *The Expression of the Emotions in Man and Animals,* published in 1872 after thirty years in the making. Darwin argued that emotions and their expression do not liberate us from the "lower" orders but link us: "The young and the old of widely different races, both with man and animals, express the same state of mind by the same movements." The way humans express emotion is universal, Darwin argued. It's part of our common heritage.

For much of the twentieth century, though, the antithesis of this argument was in favor: facial expressions were regarded as products of culture, not evolution; like language, they were seen as a peculiar reflection of each society. There was nothing innate about a smile or a pout or a grimace.

A century after the publication of *The Expression of the Emotions in Man and Animals*, a young researcher named Paul Ekman proved, contrary to his own expectations, that Darwin was right. To find a people whose conception of emotional expressions hadn't been reshaped by movies or television, Ekman traveled to the jungles of Papua New Guinea, where he lived for several months with a hunter-gatherer tribe. When Ekman, through a translator, read its members a brief story and then presented pictures of different facial expressions, asking his listeners to select the expression that went with the story, they picked, the vast majority of the time, what Ekman—what we—would have picked, too. After some academic tussling and subsequent studies, Ekman's conclusion was engraved in granite: the recognition of facial expressions is a product of evolution, not culture. I see a smile, you see a smile, the Foré people of Papua New Guinea see a smile.

Smiles weren't born yesterday, in other words. There's an epic evolutionary backstory to the upturned corners of a baby's mouth. Smiles came down from the trees, migrated to the far corners of the earth, and surfaced again in the bassinet by your bed.

That's how the smile gets to the baby, at least. But what does the baby get from the smile?

8

Get Happy

The zygomatic major is the essential tissue of civil society. A thin, unheralded strip of muscle, it is tucked just beneath the skin on each side of your face, slanting upward from the corners of the mouth toward the cheekbone. It's connected to the brain stem via the facial nerve, the public transport service for all facial expressions, its central depot a group of motor neurons known as the facial nucleus. When the zygomatic major muscle contracts, acting on a signal from the facial nerve, your lips stretch upward and outward. You smile.

No one, not novelists, not neurologists, knows where good feelings come from. There isn't a single neural structure that is the wellspring of positive emotion. There also isn't a single neural route; there's no Silk Road of happiness running across the brain. Various pathways have been uncovered and there are theories about others. The reason for the smile determines its route: a deliberate smile and a spontaneous smile originate from different places in the brain, even though they sometimes show up looking exactly the same. (Spontaneous smiles are far more common in infancy—an infant hasn't mastered the class picture smile yet.)

There likely isn't a single neural circuit for happiness—the sort of thing that would exist if a baby's smile was a pure reflection of inner emotion. Instead, positive emotion seems to have at least a handful of neural processes working on its behalf, always on call.

The human face is more flexible than even Jim Carrey knows. Its muscles can perform forty-three distinct movements; these movements can then be combined into some three thousand meaningful combinations. (The number of theoretical combinations—screwed-up, meaningless faces—is closer to ten thousand.) The manual for the most common system for categorizing facial expressions—the Facial Action Coding System, or FACS—runs to over a thousand pages, with Olympic figure skating–level instructions for scoring. (Here's a sample from the guidelines for grading a raised inner eyebrow: "Inner corners of brows raised *slightly* manifest by hair moving or evidence of muscle bulge developing, showing that the inner corner area has been pulled up. In some people the eyebrows will not move but the skin above them will move upwards." And so on.)

For smiles, the most significant combinations of movements have to do with the lifting of the cheeks, the wrinkling of the skin by the eyes, the dropping of the jaw. But smiles remake every feature. Darwin, recalling his son's first smiles, noted that "it was remarkable how his eyes brightened whenever he smiled, and his nose became at the same time transversely wrinkled." In an invisible instant, as Darwin saw, a smile can resculpt the whole landscape of the face.

The Paradox of the Joylessly Joyful Smile

In his research on emotion, Darwin used the French neurologist Duchenne de Boulogne as a primary source, and even today any analysis of a smile owes a substantial debt to Duchenne. An extraordinary, eccentric scientist, Duchenne doggedly tried to trace the entire musculature of the face. For Duchenne, this was not mere physiology: if he succeeded, he believed, he would have a map of all possible emotion, a key to the stirrings of the soul, made visible by the muscle fibers beneath the skin.

To assign specific muscles to specific movements, Duchenne used electrodes to isolate and contract each muscle in a model's face. He then snapped a photograph, capturing a permanent record of the expression. Duchenne's book on his research, first published in 1862, reproduced nearly a hundred of these photographs. They form a macabre, bewitching gallery, a work of serious medical science that today looks more like a freak show. The electrical stimulation was apparently painless, but the expressions it produced often *look* extremely painful. Many of the photographs, in which the electrodes themselves are sometimes still held up to the face, make for uncomfortable viewing.

But the expression Duchenne is associated with today is not that of pain or fear or anger. It is the smile. And not just any smile—the smile of genuine enjoyment. The upward curve of the zygomatic muscle alone often appears insincere, Duchenne realized. A true smile is signaled by the combination of the zygomatic muscle and the orbicularis oculi—the muscle that lifts the cheeks and folds together the skin at the edge of the eyes. The true smile isn't in the mouth. It's in the eyes. The orbicularis oculi, Duchenne wrote, "does not obey the will; it is only brought into play by a genuine feeling, by an agreeable emotion. Its inertia in smiling unmasks a false friend."

Duchenne's models, who were mostly his patients, were not especially attractive. He took their appearance as a challenge: he used plain subjects, he said, because the true representation of emotion—his goal—would make anyone beautiful. His photograph of the eye-crinkled smile is of a balding old man, whose remaining hair is trying to make a run for it. His toothless mouth is wide with a grin; his eyes are tightened and the skin around them strained, fanning outward.

There is a glimpse of genuine pleasure there. Darwin showed this photograph and a different photo Duchenne took of a smile—this time with the mouth alone smiling—to a number of people and asked which photo represented a smile of true happiness. All of them picked the former. This particular configuration of facial muscles, later dubbed the Duchenne smile, is considered the smile of joy. That's a loose definition, and there are exceptions and complications, but think of it like this: it is the smile that typically flashes across a baby's face when her mother approaches.

But when I look at Duchenne's photograph of it, I see not joy but irony. The photo has all the musculature right. But it has the emotion wrong. I can't see it. It's a reminder, yet again, that a smile is a slippery thing: the very first scientific documentation of *the* joyful smile fails to be joyful.

By the time we hit adulthood, the muscles in the human face combine to form a bewildering variety of smiles, and most of them have nothing to do with joy. There are dominant smiles and submissive smiles, the politic smiles of good graces, the slightly awkward, slightly hostile smile of meeting your ex-girlfriend's boyfriend. Paul Ekman has catalogued eighteen types of smiles; he estimates there might be more like fifty.

The sheer number of smiles is only half the story. The rest is how culture dictates facial expressions: this is when you laugh, this is when you cry, this is when you hold it in. The complexities of these display rules render virtually every expression suspect. The examples are endless. Show an American and a Japanese subject a disturbing movie and when each is alone, they react the same way—disturbed. But when the Japanese subject watches the movie with someone else, he will, unlike the American, smile to cover up the distress he'd displayed before. And culture alters not just the formation of facial expressions but the perception. When deciphering an expression, East Asians look at the eyes for clues,

Westerners the mouth. There is emoticon-based—yes, *emoticon*—evidence for this: in the American version of happy and sad emoticons, the mouth changes. In the Japanese version, the *eyes* change. The Japanese smiley face—read from top to bottom, not sideways—looks like this: ^_^; the sad face like this: ;_;.

Ekman established that facial expressions have a universal basis. But culture can obscure expressions way past the point of intelligibility. Babies provide a path out of this morass. Since they don't follow display rules, at least not immediately, they offer the promise of getting closer to whatever a smile, in its purest form, is actually about. The problem, of course, is that babies can't confirm what any expression is about: they stubbornly refuse to explain what they are feeling. That's why facial expressions never bear as much weight as they do in infancy. With babies, they're all we have to go by.

The Importance of Being Emotional

When I began wondering about how babies work, I had a few very basic questions in mind. I thought this meant they'd be easy to answer. I was wrong: it meant they were harder. This is why:

1. Babies have a way of turning basic questions into big questions, and sometimes even bigger questions, and sometimes into the biggest questions you can ask. The study of infancy is often about the future—how the first few years shape the rest of life. But it is also about the deep past. When scientists do silly experiments comparing baby humans with baby chimpanzees, they are ultimately asking about what separates us from the rest of the apes. Infancy research is an excavation of our foundations. The most minor questions about how babies work ultimately take you down to the sub-basement level.

2. Once you start thinking that somewhere in my baby are *the secrets of the species*, everything your child does is much more interesting for at least a few hours.

3. The bigger the questions, the more elusive the answers. This is where the structural analogy falls apart: in houses, foundations are always solid. Termites chew through walls, but the cement beneath is cement. With babies, the closer you get to the foundation, the *less* solid things seem.

Start out asking what a facial expression means for a baby and you end up asking a much bigger, nearly unanswerable question: *What's an emotion?* There isn't a settled answer to this question for adults, let alone babies. According to a recent article in *Emotion Review*, a place where they presumably spend a lot of time thinking about emotions: "Despite researchers' best efforts, there are still major disagreements about the nature of emotion (what is emotion?) and its development." Even articles in *Emotion Review* punt on the question of what emotion is.

Whatever emotion turns out to be, we know that infancy vibrates with it. This is what was so plain to Trevarthen and other researchers in the 1970s. The proto-conversations they transcribed were conducted not in words but in sentiment. With wonderful circularity, the back-and-forth between a parent and a baby is a conversation in which the content is the fact of the conversation itself: the desire to have a conversation.

From an evolutionary perspective, such a strong early motivation to engage emotionally suggests that we give too much credit to language. We want to communicate long before we have the abstract means to do so. "The impetus for language," as Sarah Hrdy has written, "has to do with wanting to 'tell' someone else what is on our minds and learn what is on theirs. The desire to psychologically connect with others had to evolve before language." She goes on: "These emotion-laden quirks of mind had to evolve before the words came along to articulate them." This is a crucial and recent insight, and a very different way to think about what made humans special: not cold logic but hot emotion. "Part of the modern revolution in thinking about the evolution of the human mind," as the evolutionary biologist Peter Ellison notes, "has come from focusing less on capacities of abstract formal reasoning and more on capacities for strategic social behavior, and even on emotion."

An early period of emotional engagement is unique to humans. The potential for it is not: when raised by humans, baby chimps excel at the sort of interpersonal skills that until recently scientists believed only humans possessed. If raised in a community that values such abilities, chimps have the capacity to develop them. But when they aren't (and in the wild, they aren't), those skills aren't nurtured.

In our hominid ancestors, they were. For once, we can't blame our upbringing. Our emotional precocity—our interest in reading and expressing emotions—evolved out of how we were raised: for a couple of million years, we looked into the many faces around us and tried to divine the mysteries therein. Humans engage with other people, Hrdy argues, because over this span of evolutionary time, "babies who were better at gauging the intentions of others and engaging them were also better at eliciting care, and hence more likely to survive into adulthood and reproduce."

So emotions are obviously crucial in infancy—crucial evolutionarily, crucial to daily life today. But how do we figure out what they are?

This is the question that separates the parents from the scientists. To any parent, the emotion expressed by a baby's smile seems obvious—maybe the only obvious thing about the baby since birth. If you polled parents on what a baby's smile means, they'd all agree on the answer: happiness. But people who study emotion and babies are nowhere near agreement. They'd give wildly different answers:

1. A smile reflects joy, without qualification, and the joy an infant feels is the same joy he will feel years later.

2. A smile reflects an inchoate, confused pleasure. It's only much later, as the brain matures, that that pleasure takes flight into joy.

3. A smile is a tool used by the infant to cultivate the relationships he'll need to survive in the world. Joy is functional: the infant learns to feel it because it is useful.

4. A smile is part of a process of feeling emotion. It isn't a sign, at least at first, of any emotion in particular.

5. A smile is just a behavioral tic, to be reinforced or erased. It doesn't have anything to do with emotion.

These are radically divergent views. You can't reconcile the idea that a smile is always a perfect reflection of joy with the idea that our sense of joy develops along with our brains. And you can't reconcile either with the idea that a smile has a strictly social function. You can hardly start a conversation between these ideas: they don't have anything to say to each other. What we parents see as obvious, in other words, scientists see as anything but.

In everyday life, we treat expressions as clues to emotions and we automatically assume that the expressions are giving us the correct clues. In fact, we go farther than that: we see past the facial expressions; we just see the emotions. "We do not see facial contortions and *make the inference* that he is feeling joy, grief, boredom," the philosopher Ludwig Wittgenstein wrote. "We describe a face immediately as sad, radiant, bored, even when we are unable to give any other description of the features."

This system works relatively well for adults, which is why we follow it, and it seems like it should work well for infants, too.

Even very young babies display what seem to be readily recognizable facial expressions.

It is stubbornly unclear, however, how meaningful these early expressions are. They might reflect deeply felt emotion; they might reflect nothing at all. Some psychologists contend that, in emotional terms, newborns are basically bipolar: they have a positive state and a negative state, and all emotions emerge, over the course of infancy, out of those primitive wells of feeling.

At the very least, there's considerable evidence that these early expressions don't always align with their corresponding emotions. Babies often fail to produce the facial expression a situation would seem to call for; sometimes they produce a wholly inappropriate expression instead: they look surprised, for example, when mouthing things they have mouthed many times before. And many infant expressions are actually "blends" of different expressions—they combine the eyebrows of someone sad, say, with the mouth of someone angry.

With so much static, it seems iffy to assume that certain expressions always match certain underlying emotions. Eventually, though, babies clearly begin to make the "right" expressions—the same expressions an adult would make. (There are many anthropological accounts of adults choosing the "wrong" expression, like grinning at a funeral, but that's culture, not development.) Facial expressions become more coherent—more correct—over the first year. Between four and twelve months, babies begin to sort out what their face is supposed to be saying. Sour foods, for example, prompt fewer expressions of joy and interest and more of disgust. Babies whose arms are gently held down show more anger. And they reserve anger for the right occasions—they don't display anger when being tickled, for instance.

An infant's skill at not just making but *reading* faces improves over time, too. Infancy is the most socially intense experience of our lives, a crush of beet-red caricatured faces. By their first birthday, babies have organized this mess into meaningful categories, at least for people they know well. Infants have faith in these categories even in a dangerous situation. The power of this sort of emotional communication has been demonstrated in an experiment on a "visual cliff," a Plexiglas table that looks as if it drops off in the middle. When a year-old baby moves from the "shallow" side to the "deep" side, it appears as if he's going to move off the edge of the table. For guidance, the baby looks to his mother. In this experiment, mothers, using only their faces—no words, sounds, gestures—posed with the expressions for either fear, anger, interest, or joy. The results were unambiguous: no babies crossed when their mothers expressed fear, and almost none when they expressed anger, but nearly all crossed for interest and joy.

Even in a situation a baby recognizes as risky, he has absolute faith in what his mother's face is saying. After a year in the world, her expressions are more meaningful to him than the peril he recognizes with his own eyes.

The Development of Joy

All thinking about cherubically smiling babies runs aground on the same fundamental problem—if you were skeptical, you might call it an insolvable problem. It's that we can never know, really, what a baby's first smile means *for the baby*. It might mean nothing. A smile might be a wonderful mistake—an expression we interpret as emotional and thus make emotional. It might have the cooperative core that evolved out of the silent bared-teeth display. It might have some other emotional resonance. It might be all of these.

The most compelling work on infant smiling takes a clever route around this immovable obstacle. It treats a baby's smile the same way we ought to treat an adult's smile: as an expression that's ultimately social. "The situation in which smiling occurs most often is social interaction," says Daniel Messinger of the University of Miami. "It's not happiness."

Social smiling is nurtured by the face-to-face interactions of early infancy—the "conversations" Trevarthen captured. At the precocious age of a couple of months, infants are already ready to live in an expressive world. They look at faces in a new way: instead of scanning the periphery of the face—the ears, the hair, the chin—they make direct eye contact. They are newly attuned to movements within the face, rather than just the fact of the face itself. They have the patience to hold someone's attention.

Out of these developments emerges the social smile. It appears when you least expect it: before smiling for the first time, a baby will stare hard at someone's face and knit his brow in deep consternation. It looks like the expression of someone preparing to frown; it looks like the expression of someone preparing to take a dump. It does not look like the expression of someone about to smile. But it is. At this early age, babies are very grave during face-to-face dialogues; they are so grave they sometimes seem serious even during their first smiles. The effort of mere interaction is exhausting. Around a month after they start smiling, they finally relax. They can play.

The smile plays a huge role in these early interpersonal exchanges: after the onset of social smiling, a baby spends about *a fifth* of all face-to-face time smiling.

There is no parity in these conversations: the baby has all the power. It is a comic imbalance. Parents can smile all they want, but the most desperate, caricatured grinning won't always elicit a smile. If the baby deigns to smile, however, the parent will almost always return the smile instantaneously. *Instantaneously* has been stop-watched here: within two seconds. Even when a baby turns a perfectly blank face toward a parent, that alone is enough to elicit a smile. Parents have no negotiating leverage: they smile at anything short of a shriek.

These dances almost always end with the baby bowing out; the parent, the sap, almost never stops smiling first. The practical effect of such a pattern, as Messinger has observed, is that a baby's "prototypical experience of smiling is smiling with another."

For a baby, this is a perfectly reinforcing system: smiles beget smiles. And those smiles are not just mirrored but amplified: a baby smiles, a mother responds with a bigger smile, the baby raises that smile with an even bigger smile—what may have first felt like a small pulse of pleasure snowballs into a feeling more like happiness. Babies like this feeling; they want more of it. With more time in the world, they smile more. The frequency of smiles, the length of smiles, the proportion of time spent smiling: they all rise with each passing month. "Social interaction," as Messinger says, "really is acting as a crucible for social development."

This is what's been fired in the crucible: between two months and six months of age, a baby looks at his mother's face for less and less time but smiles far more often when he does. He's not just more willing to reciprocate his mother's smiles. Increasingly, he's willing to be the one who smiles first.

Charting all this requires a lot of time and a lot of video footage, and if there's anyone who ought to be dead to the charm of a baby's smile, it is Daniel Messinger. Working with his advisor, Alan Fogel, Messinger once took a group of newborns and their mothers and monitored, every week and in minute detail, how and when they all smiled, up until the babies were a half year old. This is maddening work: in smiling studies, no square inch of the smile escapes scrutiny. Using coding systems like FACS, researchers measure every variable there is to measure: if the muscle around the eye contracts and whether the jaw drops open; where the infant looks, for what reason, for how long; the length and the frequency of smiles. The researchers are trying to figure out what type of smile occurs when and for what reasons. Messinger doesn't want to net a smile and pin it like a butterfly in a shadow box and then study it. He wants to see how it behaves in the wild.

If you observe these smiles over time, the way his laboratory does, you can see that certain smiles tend to show up in certain situations. Babies don't have eighteen different smiles yet, but they have clearly distinct varieties. These can be catalogued into the most adorable taxonomy ever: a taxonomy of infant smiles.

The most fundamental is the simple smile: a closed, upturned mouth, a pulse of the zygomatic major. In infancy, simple smiles can reflect positive emotion—for adults, they often reflect just politesse—but they almost always reflect a desire to *engage*: they are what infants offer strangers. Simple smiles keep what might be called the lines of play open. They express not necessarily positive emotion but the desire to *have* positive emotion. Over time, culture and experience bleach this meaning from the expression: Isaiah's simple smiles advertised his willingness to engage with others; my simple smiles advertise my willingness to grudgingly acknowledge the existence of others. His version becomes mine with distressing speed. The innocent simple smile of infancy—the expression that says *I'm here and I'm ready*—already has disclaimers attached by toddlerhood.

If the zygomatic major works a tiny bit harder, the simple smile turns into—and these are the actual technical terms—a strong smile. Zygomatic contraction appears to correlate with pleasure—the bigger a smile's curvature, the more likely there's some genuinely positive emotion behind it; the less likely the smile's fake. You can make an evolutionary argument for this correlation: the more actual effort an expression requires, the more costly it is, and the less likely faking it is going to be worthwhile.

The relationship between zygomatic contraction and emotion is do-it-at-home science: just pick a game with a really obvious climax—psychologists like peek-a-boo. (They also do a fair amount of tickling. In fact, they seem to have inadvertently proved—and this is news you can use—that tickling is just more *exciting* than peek-a-boo: "The duration of all smiles without regard to type of smile was higher in tickle compared to peekaboo games.") The high point of the game—the *boo*—will always correspond to the strongest contraction of the zygomatic major: it garners the biggest grin.

Take a simple smile and drop the jaw and you end up with a play smile. If we return to the trees, you can spot these smiles on chimpanzees during bouts of play—when baby chimps pretend to bite their mothers, for instance. This sort of smile may be the primeval origins of laughter; it shows up in the same silly, social circumstances. Back in the nursery, play smiles appear at moments that are especially, unbearably exciting. This is *all too much,* these smiles seem to say. It's as if the excitement swamps the infant, who has to open her mouth to let some spill out.

The last major type of smile is the Duchenne—the expression worn by Duchenne's doddering model and the prototypical smile of joy. It's what we picture when we picture a baby smiling: mouth like a Cheshire cat, cheeks raised to the breaking point, outer eye crinkled like phyllo. As Duchenne noted, the key muscle in the joy smile is the orbicularis oculi—it does the eye-crinkling. This very same muscle is what makes a crying baby appear not distressed but *really* distressed. It's as though the muscle is a measure of intensity, both positive and negative.

If simple smiles are a warm-up exercise, Duchenne smiles are the ball game. In the words of the Hokey Pokey, they're when a baby puts her whole self in: she bubbles over with babbling; she locks in on her partner. The physical parameters of the Duchenne smile may explain why this focus is so intense: because a baby's eyes are constricted, her vision narrowed by her crinkled eyes, she can't look elsewhere. She *has* to be zeroed in. A "duplay" smile is this plus a dropped jaw—the smile you see when the threat of tickling turns into a tickle, when *imagetchoo* turns into *igotchoo*. Its appearance is pretty much the ne plus ultra of parenthood.

A few variables—an upturned mouth, a dropped jaw, raised

cheeks, tightened eyes—are shuffled and reshuffled in almost any infant smile. Sometimes these elements are isolated; sometimes they are combined. And sometimes it is the sequence of smiles that tells the story: Duchenne smiles are often preceded by simple smiles, and when they are, the Duchenne smiles last a particularly long time—longer than when Duchenne smiles appear without any warning. The simple smiles are the ramp a baby crawls up to get to the heady heights of the Duchenne smile.

The fact that babies improve at reading and making facial expressions suggests that they have to *learn* about how expressions work. They may also have to learn about how emotions work.

Thinking about smiles as things that *develop* suggests that their deeper emotional significance—what we assume is the meaning of a baby's very first smile—arrives over time. It isn't there with the sixteenth smile; it's there with the sixteen thousandth smile.

Sometimes you can see this happening. At about three months of age, babies, for the first time, respond to the smiles of their mothers with a specific expression: a duplay smile—mouth open, cheeks lifted. It's the smile no parent can overlook. Over the next three months, babies are more and more likely to use this expression in response to maternal smiles. They know their mother wants to play; they're showing that they want to play, too. And they are less and less likely to use this expression at other times—they reserve it for a beaming mother. The increasing presence of the duplay smile at this precise moment, and the decreasing presence of it the rest of the time, suggests that a baby is coming into joy.

At the very same time babies feel this emotion more intensely, they start to regulate it—they control how much they let in. At three months old, babies are able to stop smiling while still looking at their parents; at six months, they often have to look away from their parents *before* they stop smiling. It's as if the emotion is now so much more intense that they have to physically escape its object. The more intense a smile is and the longer it lasts, the more often a baby will avert his gaze before ending it. The joy is just too much to take.

"I really think that babies are learning what joy is by sharing it with someone else," Messinger says. Smiling isn't an expression of a preexisting state so much as a path we take to *get* to that state. I'd always thought of smiling as a window into Isaiah's world. But he'd been using it to climb into mine.

So let's stipulate that emotion develops. But then, who are we *before* emotion develops? Who are we before joy?

I stammered out some version of this question to Daniel Messinger—how thinking of emotions as things that *develop* can seem like an alien idea.

He paused. "Well, you have a baby, right?"

"Yeah."

"So what's your experience . . ." He laughed.

I laughed, too, because, well, *yeah*: when you are actually with a baby, such a perspective makes perfect sense. Isaiah had gone from a spectral presence to mischievous sprite, and it was his developing emotional life, as much as anything else, that transformed him: his new joy, his anger, even his sadness, they were what gave him vitality. He'd expanded into himself, like a collapsed sack taking flight as a hot air balloon.

All that said:

Who wants to be the asshole who says a baby's first smile isn't a Technicolor rainbow of happiness? "We have a strong propensity to see those smiles as reflective of positive emotion," Messinger says, dryly. "It makes us feel connected, engaged, motivated to take care of the baby."

This is common sense and neurological fact. When first-time mothers are shown pictures of their babies, the brains of the mothers—specifically, the "reward-processing" centers of their brains—release the neurotransmitter dopamine. An addictive drug has roughly the same effect. But what controls how much dopamine is released is the expression of the babies: sad faces release a trickle, blank faces slightly more. Smiles: a flood. In the very way the brain is wired, parents are addicted to seeing their babies smile. No wonder we see so much meaning in their smiles.

In other words: to say we have a strong propensity to see smiles as reflective of happiness is an understatement. (Which is why right now you're shouting *Of course my baby's smiling because she's happy, you dope.*) But that propensity means we may be missing something even more profound: the idea that joy is an emotion created by a parent and infant together—and that parenting is ultimately about the process of imbuing a child with the capacity to fully feel such emotions.

But what happens if you can't do that—because you don't have the capacity to feel those emotions yourself? There have been extensive studies on the infants of clinically depressed mothers. (This isn't a passing postpartum depression, but something more prolonged.) When depressed mothers interact with their babies, they appear flat and withdrawn; they smile much less, for example. Their affect is so different that these babies inhabit "an atypical emotional environment," a heartbreaking turn of phrase: it means that they see far more sad and blank faces than other babies do. At five months, these babies, unlike other babies, do not detect the difference between a smile and a neutral expression, most likely because they simply are not used to seeing smiles. Not surprisingly, they smile far less themselves—they have been trained, unintentionally, to smile less.

Sometimes the training is intentional: remember the Nso mothers in Cameroon, who care most about keeping their babies calm? The Nso aren't that unusual. Gusii mothers in Kenya don't mug for smiles either; they don't even make much eye contact with their babies—they think it would be too exciting. The Gusii, in other words, look like disastrous parents. "From the American standpoint," as a couple of developmental psychologists rightly point out, "these behaviors suggest such low maternal responsiveness and engagement as to be considered possibly pathological." The paradox, they add, is that although their behavior appears pathological to us, the Gusii mothers turn out to be very responsive to their infants. They don't look like good mothers. But they are.

The converse of all this is true, too: babies can be trained to smile *more*.

Pretty much every grandparent tries to do this, of course—no one sticks out his tongue at a baby to get her to *frown*—but the Baganda, a Bantu people living in Uganda, are phenomenally successful at it. From birth the Baganda nurture smiling: they encourage their babies to smile as much as possible and to delight in the smiles of others. American mothers remember the first step; Baganda mothers remember the first smile. In ethnographic accounts of the Baganda, the smiling reaches farcical proportions:

> If a person has been unsuccessful in his attempts to get an infant to smile, he will usually ask the mother what is wrong with the child. When the mothers in our most recent study were asked what pleased them about their infants, ten of the twelve mentioned smiling. The early "sociability" of the infants in our sample was often exasperating when during their monthly examinations, they insisted on looking at the people present rather than at the test objects. Frequently, the infant had to be seated facing away from any people in order to proceed with the testing—even the tester had to remove herself from view.

The Baganda mold their children into highly social beings. It's not much of a leap to conclude that they're also molding them to take deep pleasure in being social.

I am poorly positioned to train Isaiah to take pleasure in social interaction. Temperamentally, I face inward. When I face outward, I face my shoes.

Isaiah did not face his shoes. And not just because he didn't even have shoes.

We first realized how social he was—how eager he was to face the world—when we turned him around in the stroller. This was not supposed to be a momentous occasion; it was just practical. Isaiah had a convertible stroller, the type where the car seat snaps in place, facing backward. It's a cruel fact of early infancy that even when you get to go outside, you still have to look at your parents.

He clearly wanted to see more of the world than us. But once he did, once he met the neighbors eye-to-eye, it was immediately evident that he hadn't learned the cultural display rules that governed our shoe-gazing, grad-student–infested neighborhood. *He didn't know he wasn't supposed to smile at people.* Pushing Isaiah down the sidewalk was like taking a walk in an alternate neighborhood. As he wheeled past strangers, their firmly set faces would dissolve. Faced with a baby smiling at you like his life depended on it, what was a person to do but grin and bear it?

This was Isaiah's golden Lab phase: he tried to befriend everyone in sight. He looked like a frog trying to catch flies—mouth open wide, eyes narrow, tongue hanging out. He had no idea what to do with the people he caught. If someone came too close, he'd wail. But if you kept your distance, he could perform a brilliant, if topically limited, charade of a dinner party conversation.

Isaiah will never again smile as much as he did during this hyper-social stretch of infancy, and like any parent, I have to stop myself from saying more about it—I want to talk about how contagious it was, how poignant. But who wants to read about that? (Talking about a smiling baby is boring in the same way as talking about a great relationship. There's no drama in contentment.)

Isaiah's eagerness to smile was puzzling in a way, though. Seen from the cold competence of maturity, it is surprising that creatures as immature as infants find so much in the world to smile *about*. If you're an infant, you can't do much, and in the unlikely event that you can do something, you're not very good at it. The rare things you can do well, like sucking on computer cords, you're not allowed to do. You're clumsy and uncoordinated. You're overwhelmed by hunger or exhaustion. You're incontinent. "Given these limitations and disruptions," the psychologist Edward Tronick once asked, "why don't infants typically fail to achieve their goals and continuously experience negative emotions?" It's a good question. If you're a baby, what's to be happy about?

Well, there's your parents.

Parents turn failure and frustration into joy. And by turning their baby toward that ecstatic emotion, they shape the contours of her emotional world. As Messinger says, "That experience—of being with babies and being involved with their emotional lives—that's real. It's real to the baby."

9

A Few Words on Some
Small Subjects Like Culture,
Civilization, and the
Origins of Happiness

All this may make parents feel like very important people. But fame is fleeting. With each passing week, infants are less and less content to spend eternity and a day staring into the eyes of their parents. They become distracted. With the wide eyes of adolescent boys who have just discovered Internet porn, infants discover *objects.*

Life is never the same.

Once other things—dogs, rubber giraffes, the coffee grinder—enter a baby's consciousness, the face-to-face intimacy that defined the few months after birth is lost forever. It is, in miniature, the first time the kid leaves for college: he has moved past his parents.

Over the first half year of life, the subject of face-to-face communication is the communication itself: the conversation is about the fact of having a conversation. Over the next year, this interpersonal foundation, laid down in hours of hard smiling, is built on. Babies begin to discuss other things entirely. It's as if they feel settled enough in their world—the world of faces—that they can explore outside it, at first tentatively, then more boldly.

Smiles continue to play a crucial role in these expeditions: they're how babies tell other people what they've found out there in the world. It's not that babies are smiling in new ways; they don't learn more complex combinations of facial muscles. Narrow eyes, contract right side of the zygomatic major only, wiggle ears in syncopation: they don't do that. What changes is when they smile and where they aim those smiles. Without language and without any new gestures, babies are able to communicate, often very clearly, about a wide new range of events and things. Before words come smiles.

The crucial element of these smiles—the sort that connect people to things—is not the zygomatic major, or the orbicularis oculi, or some yet more esoteric part of facial physiognomy. It's something very basic: the eyes.

Continuously, unconsciously, we follow the eyes of everyone we meet, tracking where their eyes are looking, trying to deduce why. It's the closest we get to telepathy. Humans aren't the only species who follow a gaze. The great apes do it; so do dogs. But we're especially good at it because we have an advantage: we are the only primate with a dark iris on top of a white eyeball (or more precisely, on top of the eyeball's white covering, the sclera). The stark contrast leaves the iris nowhere to hide: its focus is apparent at a glance. (By a year, it is apparent to babies, too—when someone's head goes in one direction and their eyes in another, the babies stay with the irises.) Other primates have hardly any white in their eyes. Their iris is indistinguishable, its movements hidden by the sclera, and the direction of their gaze is effectively disguised. Our primate relatives would be really good at covertly ogling someone; we are indisputably bad at it and we always will be.

In a sense, we evolved to be bad at it. The unique design of our eyes may have enabled the uniquely human act of cooperation.

Before language arrived on the scene, being able to read the direction of a gaze may have made planning and coordination possible. Without this sort of collaboration, human society, a deeply interpersonal enterprise, is inconceivable.

It is through the eyes that babies elaborate face-to-face smiling exchanges into more sophisticated communication. They coordinate gazes with other people about other things—you *and* I are looking at *that* fire truck. Sharing a common subject with someone else is called "joint attention" and it marks a developmental tipping point.

It is worth pausing here to note that joint attention is *a very big deal,* in ways that have to do with the evolution of the species as well as the development of the individual. Joint attention is, for example, central to the development of language: it is how infants acquire a vocabulary—it is how they learn the words "fire truck." It may even account for the evolution of language: without joint attention, you don't have two people paying attention to the same thing; without the same thing in mind, you don't have much chance of coming up with a language.

The choreography here can be confusing: if a baby happens to look at a toy that someone else is looking at, that's not joint attention—that's just happenstance. A baby has to be aware that the other person is paying attention to the object. Ideally, as the psychologist Peter Hobson says, "he or she should share awareness of the sharing of the focus"—the baby should know that you're both looking at the object. This typically involves having the same attitude toward whatever you're looking at. As in, both of you agree that that fire truck looks *awesome*. Ultimately, joint attention is about being able to share or at least coordinate your feelings with those of someone else. To join with someone else in thinking about *something* else is to be able to sense that other person in yourself, Hobson argues. This is the crucial leap that children with autism are largely unable to make.

At around a year of age, babies experience, as the psychologist

Michael Tomasello puts it, "a revolution in their understanding of persons." They begin to realize that other people are beings whose actions are goal-oriented—that these mysterious beings act according to what they're trying to do. Babies begin to understand that people have *intentions*.

Once a baby understands the idea of intentions, he can share not just the same focus but the same goal or task. And once he develops this capacity, it is manifest in almost every activity: a baby and his father mock feeding each other, or rolling a ball back and forth. It's so crucial a capacity that it is hard to see how the same baby, only a few months before, got along without it. Without this understanding, humans would achieve nothing at all—our daily lives not only depend on it, they assume it. Evolutionarily, it is fundamental to who we are: only human beings work together toward shared goals.

Tomasello, who conducts experiments on babies and on other primates, is a hugely influential thinker on this subject, and his theory-of-everything argument goes well beyond language. He believes that the gap between us and our primate relatives was wedged open by factors that were social. It's not that we had more per primate brainpower, he says. It's that we were able to collectively create a culture that passed on its practices and innovations, allowing a "ratcheting up of cultural and cognitive complexity" over time. And what lies underneath all that are the fundamental capacities you can see emerging in infancy: the ability to meld minds with someone, to act toward a common goal, to feel a shared emotion. When babies and adults work toward a shared goal, they are engaging not just in "cultural learning," Tomasello says. "They are engaging in full-blooded cultural creation."

This is a grand idea that is surprisingly easy to domesticate, if you are willing to be undignified enough: call it the premasticated pear theory of everything. This theory states that all human institutions, when you really get down to it, are built on a single timeless impulse: the desire of babies to offer you handfuls of mushy pear. These generous, sloppy gifts reflect a capacity to share a subject (the pear), a goal (eating the pear), even an emotion (the pear makes me feel good). It's an act of complex, interpersonal engagement, and it happens a dozen times at every meal. Within this act are the capacities for cooperation, the foundation for how we got together and figured out other things to do with pears. It's how we figured out how to plant pear orchards, or poach pears in red wine, or make pear eau de vie—a heady step up from a handful of mushy pear.

So Happy Together

After spending six months in the world, babies, regardless of their prenatal exposure to Tomasello's work, grasp the importance of eyes. They track a gaze as it drifts across the room. They connect their old love (bedraggled parent) to their new love (neon teething ring) by bouncing their gaze back and forth between the two. In the weeks and months to come, they *initiate* joint attention by directing their parent's attention toward the teething ring. Soon afterward, babies layer another element on top: they combine all this with a smile.

These new smiles aren't transactional. They are less likely to occur when babies are trying to get something from someone— when they're requesting, or demanding, something. Instead, they appear when an infant is showing off a toy or trying to share yet more premasticated pear. These sorts of smiles aren't meant for wheedling more Cheerios out of the parents, at least not initially. They're a way to comment on the world—and how the world makes you feel.

Twenty years ago, Susan Jones, a professor of psychology at Indiana University, constructed what she called "an infant analogue to a bowling alley," a riff on the famous study of bowlers not smiling. The "bowlers" in this scenario were ten-month-olds; the "bowling balls" were a few toys, affixed to a wall. Mothers sat in a chair a few meters away, facing the wall—roughly where the rest of the bowling team would be. Since the toys were attached to the wall, the infants couldn't play with their mothers and the toys at the same time. They had to negotiate between the two.

Some mothers were attentive; some mothers were inattentive. But it turned out that whether they were attentive or inattentive affected almost nothing about how their children played: it had "no effect on the infants' total time in play, the persistent quality of their play, their tendency to smile at the toys, or the timing and lengths of their looks at mother." The only thing it affected was the frequency "with which infants smiled in the mothers' direction." The infants figured out whether or not their mothers were likely to be paying attention and awarded their smiles accordingly: mothers who were paying attention got lots of smiles; mothers who weren't got fewer. If the mother wasn't paying attention but a stranger nearby was, a baby would aim the smiles at the stranger instead. Such "an audience effect," Jones concluded, suggests that infants use smiles as social signals. They smile for someone.

The "infant bowling alley" experiment identified a new type of smile: the anticipatory smile, when a baby smiles at an object and then looks at a person *while maintaining that smile.* He's had an emotional response to something and wants to share that response with someone else. There's a very satisfied quality to these smiles: they seem to say, Whoa—that was *fun.* Too much fun to keep entirely to yourself, especially if you know the other person is also going to think it was fun. That's what anticipatory smiles are built on: the baby's knowledge that the other person is going to share what he's thinking. That's what makes this smile—so seemingly routine you might not even notice it—so wonderful.

"When the baby smiles and then shares that smile with the other person, it's almost as if they're showing us that they've learned all they needed to learn from face-to-face—that is, they know the other person is likely to also be positively disposed to-wards the object and towards them," Messinger says.

There's growing evidence that anticipatory smiles mark a cru-cial moment in development: the moment when babies begin to intentionally communicate about their social world. The same infants who form anticipatory smiles act in other ways that are clearly intentional—they gesture or babble in a way that's more obviously communicative, or they use toys toward a particular end, showing that they're capable of means-ends behavior (if I do this thing, that thing happens). It's a combination of behaviors suggesting that when babies smile and then aim the smile at someone, their message is very much intentional and meaningful.

The timing of this smile is everything. Imagine a baby looking at a waddling pull toy of a duck, looking at an adult, and only *then* smiling. This smile—a reactive smile—might be more about the adult than the duck. You certainly can't conclude that the infant was feeling swell *before* he saw the adult.

Unlike these reactive smiles, or even the desire to share some-one's attention about something more generally, the number of anticipatory smiles rises steadily between about eight months of age, when they first show up, and a year. It's a sequence—smile at object, hold smile, smile at person—that's entirely new: at six months of age, it happens less often than it would by chance. When it appears a couple of months later, it is there for a reason.

As Isaiah approached his first birthday, his total smiling had shrunk: where he'd once been profligate with his smiles, he was now penurious; where he'd been eager, he was hesitant. He'd gone from an easy mark to a tough room.

But at the very same time, he was flashing more and more of these anticipatory smiles, although I didn't know then that they were anticipatory smiles. With typical fatherly narcissism—the yeah-well-my-child-*invented*-childhood variety—I thought they were unique to him. A lot of these smiles involved the pink-pig-with-wings weathervane down the street, which did seem like the sort of experience you'd want to share with someone else. I did my paternal best to confirm that I'd also seen the pig with wings and that it was indeed marvelous—although my pleasure was complicated by the knowledge that, at some highly hypothetical point in the future, we still needed to get to the corner store.

An anticipatory smile doesn't reflect a desire to share just any experience. It reflects a desire to share a *positive* experience (which typically—but not always—involves a pig-with-wings weathervane). And somewhat amazingly, a baby's level of anticipatory smiling—her interest in sharing positive experiences—seems to predict future social behavior: levels of anticipatory smiling at nine months of age successfully predict social expressivity and competence—basically, the ability and inclination to engage with other people—at thirty months, almost two years later. (Levels of reactive smiling, in contrast, don't predict anything at all.) "Kids who want to share positive things about the world are perceived as more socially competent—not getting into tussles, playing well with others," Messinger says.

Smiling is a social high point that motivates babies to seek out *more* smiles, he argues, and you can see that borne out in these studies. Babies aren't born with the goal of mutual smiling: they develop that goal. "You're really looking at early feelings that are likely to play a motivational and regulatory role in people's lives," he says.

That's what's so poignant about anticipatory smiles: they suggest that positive emotion—the welling up of joy—is what motivates us to communicate with other people about the world. The experience of smiling prepares us to find pleasure in being with others—and in the long run, to participate in society. As Messinger has written, "Affecting and being affected by the positive emotional expression of the parent may lead infants to experience the happiness of others as essential to their own happiness." It's a powerful idea and it is powerfully moving—by the standards of developmental psychology, it is almost a sonnet. Isaiah didn't smile at me because he was happy. He smiled because I was happy. And, in due time, that was what taught him to be happy, too.

Part Three

TOUCH

10. *Handle infant as little as possible. Do not pick up to show to relatives and friends. Keep quiet and free from commotion and infection.*

13. *Do not pick baby up every time it cries. Normal infants cry some every day to obtain exercise. Infant is quickly spoiled by handling.*

19. *Baby should never sleep with you or any other person.*

24. *Don't listen to the careless advice of friends and relatives. Do as your physician advises. He knows more about you and your baby than they do.*

—*Instructions to Mothers* brochure,
Research Hospital of Kansas City, 1949

And I kept to the strict rigid routine of not cuddling her when I wanted to—you fed her, changed her, winded her and put her in her pram, and pushed her up to the end of the garden and that was it. And you really didn't give the cuddles and the love and affection. If I kissed her it was on the back of the neck and nobody else was allowed to kiss her. That was something that we did bar, was kissing. I kissed her little body and I kissed her neck but never kissing her as such . . . I wasn't a tough person, that's the funny part about it all, I wasn't hard and I wasn't tough and I wanted to love my baby, but this regime was very much against my desires and I'm sure loads of other mothers must have done and felt the same thing.

> —Edith Broadway, an English mother, on raising her
> daughter (born 1935) according to the popular
> baby manuals of Truby King

On a trip to Kenya in 2004, I monitored an exchange of letters and op-ed pieces in the Nairobi Standard *concerning the growing popularity of baby strollers. All of the many medical and child care authorities contributing to the debate condemned the contraptions as devices of torture that forced children to be separated from their mothers' loving touch.*

> —Nina Jablonski, *Skin: A Natural History*

10

The Power of Touch

Saul Schanberg wasn't looking for what he found. A neurologist at Duke University's medical school, Schanberg had been analyzing the molecular systems in brain development when he noticed an odd correlation. When a rat pup was removed from his mother, his biochemical functioning would go haywire. The pup's levels of growth hormone and of an enzyme crucial to tissue growth and the immune system plummeted. Once a pup was returned to his mother, the levels snapped back to normal.

Schanberg and his lab puzzled over what might seem like an unanswerable question: What *exactly* did the mother provide that was so critical? Without the mother, the rat pups were unable to develop properly: the whole physiological superstructure that undergirds normal development went wobbly. But why?

Schanberg systemically ruled out a series of potential causes. Feeding was not it. Nor was being able to see or hear or smell the mother. "The question then arose," Schanberg later wrote, "as to what it was that mothers normally do with their pups, other than feeding, emitting sounds and smells, and so on, that is then missing during separation."

An investigation of the molecular systems of brain development had ascended, improbably, to almost psychoanalytical status. How did the rat pups know their mother was around? At the most elemental level, what was a mother good for? Why did she *matter*? After months of failed hypotheses, a graduate student of Schanberg's walked out of the lab and returned with a paintbrush. Dipping the brush in water, he firmly stroked the pups. Their biochemical imbalance righted itself: the pups read the paintbrush as their mother's tongue.

It turned out that the question—*what's a mother for*—only seemed unanswerable. Reduced to its core, Schanberg's experiment had an unexpectedly, poignantly simple answer: touch. A rat pup cannot live without a mother's touch. Or failing that: a paintbrush.

It is impossible to isolate the influence of touch in a human pup. Studies of infants who were severely harmed by deprivation—who had been in the infamous Romanian orphanages, for example, where infants received the bare minimum of human contact—can rarely pinpoint what caused the harm. All the variables are confounded, as the scientists say: you can't tease out the different toxic components of deprivation and control for them. And you certainly can't design an experiment that does so (although before modern ethical standards, as recently as the 1940s, some psychologists tried).

But there are many good reasons to think that Schanberg's answer applies to us, too—that we are all, in a way, rat pups. Rats and humans share roughly the same neural systems. Rat pups are born premature, like humans. They require extensive social interaction—like touching—to properly guide their development. Rats are devoted, highly motivated mothers.

But—and this is the weird part—all that overlap is almost unnecessary. The obvious reply to a sentence like *if you're a rat pup, you can't live without touch* is to say, *but I'm* not *a rat pup*. I'm sure there's no shortage of experiments suggesting connections between rats and humans to which I'd immediately reply, *but I'm* not *a rat pup*. But in this instance, I'm immediately convinced that I *am* a rat pup. All the neural overlap hardly matters. Even the species hardly matters: it could be lesser anteaters. I already, whether for rat pups or the offspring of lesser anteaters, believe that touch is life itself.

Touch is life itself. It's a deep, swelling feeling, but in print, it looks like an obnoxious, passive-aggressive advertising slogan. What does it mean? I suppose—and touch is an amorphous subject: it involves a lot of *supposing*—that in infancy, touch is where meaning is found. All the crucial messages sent by a parent—love, security, commitment—are communicated through touch. It's our lifeline.

Babies swim in a world of touch. Much like Moliere's Monsieur Jourdain, who discovers he's been speaking prose his whole life without knowing it, a baby has been *feeling* without ever realizing it. Since birth, he has lived in a world demarcated by skin and nerve endings. He's been swaddled and kissed, poked and pierced, fondled and, quite possibly, circumcised.

A newborn cannot reach out and make contact with the world: he may have the desire, but he doesn't have the execution—he's shaky in his movements, erring in his aim. For a newborn, the sense of touch is about *being touched*. The baby doesn't go to the world; the world comes to the baby. And the desire to grab a baby and hold him close—to press him against your chest—is so strong and sweeping it can be overwhelming. Anya had a very real fear, expressed many times before Isaiah was born, that she would squeeze him to death. This anxiety came up at a dinner party once, where it turned out that a friend's sister, as a child, had in fact squeezed her pet hamster to death. This was not reassuring, especially when Isaiah turned out to be not disproportionately larger than a pet hamster.

A classic experiment: a few hours after birth, mothers correctly identified their newborns with no information other than what the back of their tiny hands felt like. That was it: *the back of their hands*. The mothers weren't able to see or hear the babies. They hadn't even been near their children since agreeing to participate in the experiment, so they couldn't have mentally registered the texture of their skin. All the knowledge the mothers had was acquired unintentionally. It was part of how they breathed in their new child.

The drive to touch—to *squeeze*—our children, and sometimes our pet hamsters, has a wave of biochemicals behind it. Touching and being touched spur the release of oxytocin and endorphins, creating a surge of well-being—of doped-up trust and pleasure. They lower stress hormone levels and increase the quantity of serotonin, the neurotransmitter at work in antidepressants. In studies of other primates, those who spend the most time grooming—touching—are the least anxious and depressed.

These biochemical effects do not need to be translated. Unlike language, touch is instantly comprehensible. Another experiment, this time from the perspective of the newborn: placed in different environments that stressed different senses, infants were most calm when the only sense they experienced was touch. Amazingly, they cried less when they were touched by their mothers (but could not see or hear them) than they did when they could see and hear their mothers *and* be touched by them. Touch alone is the deepest form of reassurance, a shield against the world. Newborns who are held skin-to-skin for fifteen minutes before a painful medical procedure cry afterward for a couple of seconds. Newborns not held skin-to-skin cry for over thirty seconds; many cry for much longer.

Touch is the root of all other senses, the medieval philosopher Thomas Aquinas thought. It is, he wrote, cleverly, "the one which entitles a living thing to be called sensitive." Discussions of touch invariably slip into this type of foundational language, perhaps because touch is simultaneously everywhere and nowhere. "Touch is at the same time the most complex and the most undifferentiated of the senses," writes the historian Sander Gilman. "Sight, hearing, smell and taste all have specific, limited sensory organs, all of which have specific limited functions." Touch does not: there's only one way to see, but there are many ways to feel. For starters: touching someone is different from someone touching you. Both are different than the sensation of proprioception, the feeling of your body in space, or kinesthesia, the feeling of your body moving through space. A sensation can be pressure or pain, searing heat or the whisper of a tickle.

All these layers of meaning multiply the sense's potential power. "The physiology of touch, much more than the other senses," Gilman says, "reinforces the potential for fantasy about its nature." This may be why touch, and the importance of touch in infancy, is a subject upon which people like to make Strong Assertions, or Very Big Pronouncements, or Deep Statements of Faith. If the story of sucking tells us the most about the parents and experts of the past, and smiling the most about the experts of the present, touching says the most about the parents of the present: for good and ill, it is our story. It's at the heart of how we express the emotional hugeness of parenthood. How do I get all this *feeling* across to this tiny thing?

It's a shock to learn that touch wasn't always the answer to this question. Until recently, in fact, it was the wrong answer. Too much touching could spoil the child; too much touching could even kill the child. Historically, people didn't spend a lot of time thinking about touch as a sense—it was so elemental it was missed entirely. But when they did think about it, they did not always think nice things.

Almost no parent thinks this way today. And they are not alone. When talking about their research, scientists act a lot like parents: just as nearly every parent will tell you his baby is the most beautiful in the world, nearly every scientist given the opportunity will tell you that his work explains how the world works. Saul Schanberg, who died a few years ago, was no exception. "Touch is ten times stronger than verbal or emotional contact," he once said, "and it affects damn near everything you do." He went on:

> If touch didn't feel good, there'd be no species, parenthood, or survival. A mother wouldn't touch her baby in the right way unless the mother felt pleasure doing it. If we didn't like the feel of touching and patting one another, we wouldn't have had sex. Those animals who did more touching instinctively produced offspring which survived, and their genes were passed on and the tendency to touch became even stronger. We forget that touch is not only basic to our species, but the key to it.

Touch is the frog prince of the senses: once disparaged, if noticed at all, it is now heir to the kingdom.

My first memory of touching Isaiah is of not touching him. Holding my son in the hospital felt like holding an abstraction—as if he were still only the idea of my son. Tiny, almost buoyant, he was lighter than he was before he existed. In my head, he'd had a weight that was crushing. In my hands, he almost fluttered.

I had no idea how to hold him. I was so conscious of not hurting him that I was only vaguely aware that I was holding him, too. Mostly what I was doing was not dropping him.

It was an ironic introduction to fatherhood. Almost all the memories that follow are of clutching. And of fondling. This is the wonderful thing: you are allowed to fondle your baby, although you are supposed to use a different word. You get to know a baby by taking inventory—the way his wizened toes curl, the Platonic bottomness of his bottom, that blue bulge of a blood vessel on his nose. A baby's body doesn't quite belong to him yet. It was made by yours; you still have a claim on it. And it is far more interesting than yours: freshly made, softer, smoother, stranger, *better*.

This is the most we will ever touch our children. All mothers soothe their newborns through touch. But between two and six months of age, the total quantity of maternal touch inexorably declines. What increases: words, facial expressions, gestures. The baby has left a world defined by touch and entered one where sensory inputs, even the soothing ones, come from every which way.

Once I stopped not dropping my child and started touching him, I felt a shiver of relief. This seems to have been my default response to the wonder of parenthood: relief. I was relieved when he was born and it was over; I was relieved when I touched him and he was there. But it was more than relief, really: it was a sense of *release*.

Then I started in with the squeezing.

In Which We Return to the Trees

Touch has what scientists call phylogenetic and ontogenetic primacy: it formed first in evolutionary time and it forms first in each individual. Language is by comparison a latter-day invention—perhaps as recent as fifty thousand years ago. But body language has been with us since before we were *Homo sapiens*. Our sense of gesture and touch may be not just a product of culture but of evolution, the culmination of neurological processes that have been passed along genetically. Scholars of "nonverbal communication" argue that we tend to fall back on gesture as a means of communication because it is evolutionary bedrock.

Evolutionary bedrock. Discussions of touch spend a lot of time down there. If you are looking for reasons why touch is as important as Saul Schanberg says—why we really are rat pups—there is no shortage of stories from our relatives. For infant chimpanzees, the world is mediated through touch: baby chimps cling to the maternal chest; they are carried and stroked by their siblings.

All this touching is felt long after the actual contact: it leaves a physiological imprint. Generosity of touch is passed down the generations: the best predictor for how much time a rhesus monkey mother will spend with her baby is how much time her mother spent with *her* when she was a baby. In some primate species, the production of antibodies is linked to the amount of physical contact in infancy. The more touch, the more grooming, the stronger the immune system. Touch is deeply embedded in primate societies—some species of primates spend as much as 20 percent of all waking hours grooming.

Humans do not, as a rule, spend a lot of time jealous of our chimpanzee relatives. We have empathy and iPhones. Once we catch insects, we can deep-fry them and add salt. But sometimes all these advantages look meager compared to the abilities of the mother chimpanzee.

Baby chimps and baby humans start out the same way: they weigh their mothers down. A mother chimpanzee clutches her newborn with one hand while scampering along on her legs and her other hand. Three-legged chairs aren't much good for sitting and three-limbed chimpanzees aren't much good at moving: new mothers frequently fall behind the group as they travel from place to place. It's the struggle of the working mother, even when the mother is a quadruped.

But any cross-species maternal empathy with chimpanzees ends here. After a couple of weeks, a baby chimp is able to hold on to her mother's belly without any assistance. This is the image we have of the baby chimpanzee: suspended, bouncing upside down, the mother swinging through the trees, unimpeded by her newly acquired cargo. It's a beautiful, balletic act of togetherness and it ought to taunt *Homo sapiens* parents: it suggests that juggling work and parenthood isn't impossible, after all—it just isn't possible for us.

We fell from this state of maternal grace for very earthy reasons: we needed to sweat. Early humans desperately needed to
stay cool, partly because their energy-consuming brains were expanding so rapidly, and by shedding body hair, they became phenomenally, embarrassingly good at sweating, thus preventing
overheating. *Shedding* isn't exactly right: even though humans are
nicknamed the naked ape, we don't technically have less hair than
our primate relatives. It's just much, much thinner, sometimes so
thin it nearly disappears. No infant could cling to this hair: there
isn't enough *there* there.

At this late point in our evolutionary divergence, though, even
the woolliest human wouldn't be any help to an infant. Chimpanzees have big toes that function like our thumbs: they grasp and
cling. But the big toes of early humans changed with bipedalism:
they adapted to the stresses of walking. Now they aren't any different from the other toes. Even if a human baby had a hairy chest
to clutch, his feet wouldn't be any good at it. But infants still have
a very resilient grasping reflex, a legacy of this evolutionary history. In fact, the reflex is strongest during crying—it reflects how
much a baby still wants, however many millions of years later, to
hold on.

He can't. He's screwed from all sides: he doesn't have anything
to cling to *and* he's lousy at clinging. He needs *you*. By the time of
Homo erectus, it seems likely that infants could no longer hold on
to their mothers. Their sudden helplessness left their mothers in a
bind: the mothers still had to collect food and water—they had to
gather enough to survive—but they now had to do so while holding on to a child who could no longer hold on for himself. This
part of the story is very modern: in effect, the mother had been
given a new job without being relieved of the old one.

From the easy portability of the infant chimpanzee, this is a
leap. It represents a radically new model of child care. "In a sense,"
as the anthropologist Dean Falk notes, "we can think of our babies'
helplessness as the result of an evolutionary 'natural experiment'

that severed prolonged physical contact between hominin mothers and infants." *Severed prolonged physical contact:* if it sounds trau-
matic, it is because so much was at stake. "For more than
thirty-five million years, primate infants stayed safe by remaining
close to their mothers day and night," the anthropologist Sarah
Hrdy has written. "To lose touch was death. This explains why,
even today, separation from a familiar caretaker provokes first
unease, then desperation, followed by rage, and finally despair."

Mammals typically solve the problem of child care in one of
two ways: they "cache" their infants somewhere safe, and return
to feed them periodically, or they carry them constantly. Almost
all primates go with the latter. Even after human infants were
"severed" from their mothers, they were almost certainly too
vulnerable and valuable to be stashed away out of sight. Instead,
they were humped around. They might have been put down
occasionally—practically speaking, they had to have been. (Dean
Falk has argued that these periods apart created a need for *vocal*
contact, which motivated the development of motherese, which
in turn sparked, somewhere way down the line, the development
of language.) Even counting these brief separations, evolutionary
history suggests that early humans stayed extremely close to their
infants nearly all the time, almost always because they were actually carrying the damn things.

An infant who has to be carried—who can no longer hold
on—is a literal burden. How much of a burden? According to a
recent experiment, it is 16 percent more of a burden to carry a
baby in your arms than in a sling. In the unforgiving calculus of
caloric expenditure, that adds up to an enormous amount more. It
outweighs even the costs of lactation, itself a massive burden.
"Would any biped be able to travel far enough and fast enough
to gather resources, escape from predators, and keep up with
her group while incurring an average of a 16% increase in cost
above the cost of the baby's mass alone?" the authors of the experiment ask. "Would such a cost be too big to allow successful repro-

duction?" Yes, they conclude: there's no way maternal bipeds could have juggled all those jobs. Of course, the cost clearly wasn't too big for successful reproduction. Since we're around today—the ultimate beneficiaries of all that successful reproduction—those early mothers somehow found a way to do it all.

This is where the baby sling comes in. A seemingly trivial object, a side event to history, the appearance of the baby sling intersects with a fundamental story in human evolution: the invention of the first tools. Early humans, instead of hunting large animals—the discredited Big Game theory of hominin life—got the majority of their food from foraging, like chimpanzees. It was the women who did the foraging. Once babies were no longer able to cling, their mothers had to forage while holding on to an infant, an almost impossible task. The baby sling solved the problem of how to keep a baby safe and gather enough food—both utter necessities. It's now thought that the infant sling was devised by the time of *Homo erectus,* beginning around 1.7 million years ago, when relatively long daily treks were commonplace. (There isn't archeological evidence: Baby Bjorns don't fossilize.) Contrary to the Big Game picture of everyday life, what early humans really needed wasn't spears. It was child care.

The invention of the sling had a couple of far-reaching consequences. The first was anatomical: by making child care more practical, the sling may have allowed mothers to give birth when their babies were at an even more immature state. This, in turn, is part of what allowed the human brain, past the barrier of the pelvis at an even earlier date, to grow to its full, extraordinary potential. The second was social: the sling was the very first thing—the very first thing *ever*—that put some space between mothers and their babies. It was a tiny amount of space, mind you: the mother and child were still in almost constant contact. It wasn't as if the *Homo erectus* mother had happened upon a state-accredited, full-time daycare with a one-to-three teacher-child ratio and low staff turnover. But the sling was still something that came between a mother and her child: it wedged open a gap. In a quiet, modest way, it was a paradigm shift in parenting.

That gap never closed back up. Over the years, it was wiggled open wider, bit by bit. Humans have gone from an environment in which physical contact was inescapable to an environment in which it is entirely escapable. This is where the real history of touch in infancy begins: at the very moment we can avoid it. In a way, the whole contradictory story of how we handle babies—from the harsh swaddling of mother Russia to the slings of hunter-gatherers, from the age of don't-spoil-the-child to contemporary attachment parenting—is the story of how we answer the question of what we, with our upright, hairless, opposable-toe-less bodies, are supposed to do with these creatures now.

Parents—or mothers, for it is still almost always mothers—are stuck with the job that used to be the baby's responsibility. And all our anxieties about how much we are or aren't in touch with our babies—from slings to strollers to day care—stem from how we handle that responsibility.

Poorly!

Some unsolicited advice: never structure your life so that after you drop off your child at his day care, you go read the literature on the importance of not dropping off your child at his day care.

Once, after Anya and I took Isaiah to day care together, during a time when I was reading a lot of grand claims about the role of touch in infancy, I told her on the way home that he shouldn't be in day care at all and I shouldn't be writing this book at all. When I finished I would have, in the very best case scenario, a book. A book! What good was a book? Would a book care for us when we were old and decrepit? A book wouldn't care for anyone but silverfish! Would the silverfish care for us? Our child certainly wouldn't, since we'd abandoned him.

She tried to pull me out of the rabbit hole, but she couldn't.

Here's the thing: intellectually, I was very skeptical of these claims. And Isaiah's day care was only part time. And he loved it. But viscerally I had no defenses. I was immersed in lousy thinking but it was lousy thinking *about my child*. The lousiness was irrelevant; the *my childness* was highly relevant. It penetrated. Note the gratuitous "part time" clarification above: I'm still responding to the claims; I'm still on the defensive. It's only part time, I insist! I was still there, I was still around, I was still *in contact*.

Touch is as insidious as any subject in infancy: it will worm its way into your parental insecurities. Even crappy, caricatured thinking—maybe especially that sort of thinking—has this power. Which is another way of saying that if the real history of touch in infancy begins at the very moment we can avoid it, that doesn't mean the avoiding comes easily.

11

A Brief History of How We Have, and Have Not, Held Our Children

When a child is born among the Efe, who live in the Congolese rain forest, the baby is gently washed in cold water and wrapped in a cloth before being passed around by the other women present. She's then passed around by the men waiting outside. Only at that point, sometimes as long as several hours after the birth, is the baby passed to the mother: this is the first time she touches her child.

In nearly all Western hospitals until a few decades ago, babies who hadn't cried at birth were swiftly spanked, as if they were naughty in the womb. The scream told the physician that the infant's lungs were unobstructed. The spanking wasn't necessary for the newborn; it was reassurance for the doctor.

How a society welcomes a newborn is a cultural Rorschach test. You can sense just how visceral the process is by how strange every way that isn't your own feels. These days, any baby-friendly hospital in the United States places the freshly delivered newborn on the mother's chest. These same hospitals used to prize authority and hygiene: not so long ago, nurses would whisk the infant away from his mother; the infant had to be examined and scrubbed.

Now they rush him *toward* his mother. It's this mother-child intimacy, supported by a reassuring scaffold of randomized studies, that we now value. It seems *right*. The Efe's ritual, which puts the community before the mother, and even the recently departed sound spanking, which put the doctor before everyone, seem deeply *other*.

Today we think of the infant as someone who's completed a journey and deserves a bit of a break. (You've come all this way! You must be exhausted! Why don't you take a seat on my bosom?) But a few hundred years ago, the newborn was regarded as incomplete, a "lump of flesh" who still needed to be molded into proper shape. The seventeenth-century midwife, after cutting the umbilical cord, went right to work:

> She smoothed the sightless eyes, shaped the nose, opened the mouth, and rubbed the jaw to make the features firm and regular. She then pulled the arms and legs to their full extension, rubbing and shaping each in turn so that the child would grow straight and tall. Finally the midwife carefully but firmly pressed the bones of the baby's skull toward the soft spot on its head, then bound the head with fillets of cloth to further draw the bones together.

In other words: the first touch a newborn feels has not always been a cuddle. In rural Nigeria, as in colonial America, molding the skull is standard practice. Haitian peasants have traditionally done the same thing: narrow noses are prized, so the newborn's nostrils are pinched repeatedly over his first few days. The connecting core belief that a baby needs to be shaped is, in a way, exactly right: radically different cultures have all sensed that even a full-term newborn isn't finished yet—like all human babies, he's come out too soon.

All this pressing and pushing and pulling seems harsh, but sometimes the first thing a newborn feels is not even human. The

Fulani people, who live throughout West Africa, have long considered it crucial that the newborn *not* be touched by someone at birth. Instead, the baby should first touch the ground—the earthen landing gives him roots in his new home. The Kung, the hunter-gatherers of the Kalahari, have traditionally given birth alone in the bush, without even anyone else there to catch the child: the newborn plops onto a small hill of grass. The Siriono people of Bolivia fluff up a pillow of ashes beneath a hammock: the newborn slips through the strings and lands a few inches deep in ash.

Sometimes newborns are sheltered from the world: the Datoga of Tanzania seclude the new mother and child in a small, dark room for several months. Sometimes the world is thrust upon them. The Jicarilla, an Apache tribe, used to pierce both ears with a thorn immediately after birth. Among the Yoruba in West Africa, a newborn is held upside down by his feet and shaken thrice.

The Yoruba practice of head-down shaking is a lot like the very Western obstetrical spanking. Sometimes it all comes around. Spanking in the delivery room only died out in the 1970s, when obstetrics became, for the very first time, sensitive to the experience of the newborn. The French physician Frederick Leboyer led a revolt against the machinery of birth: he claimed that standard hospital practices amounted to "the torture of an innocent." Babies should be babied, not spanked, he said. At the time, Leboyer looked like a fad; today, he looks like someone saying the obvious. He just seems *right*.

Some of these traditional rituals—the ear-piercing, for example—have already been assimilated out of existence. What has not died out, though, is what may be the most ancient and oddest tradition of all. But we'll only recognize its oddness after we abandon it—only then will it just seem *wrong*.

The Political Philosophy of Swaddling

Like the baby Jesus, his Jewish forefather, Isaiah was wrapped in swaddling clothes. The precise details of the baby Jesus' swaddling clothes were not recorded but they were probably thin strips of linen, crossed from head to toe. Isaiah had a white blanket with vertical pink and blue stripes, which, in proper swaddling position, made it appear as if he'd arrived at his birth wearing a snappy V-neck T-shirt—he'd dressed to go boating but ended up being born instead. This sailing outfit has become de rigueur apparel for any newborn these days: not just for every baby on Isaiah's floor but for the babies of friends born across the country, all of whom appear on the Internet in the same neonatal wrapping. It is the fate of Isaiah's entire generation to begin their lives in gender-hedged, nautical-themed mummification.

William Buchan, had he been forced to sit through the Flickr slideshows, would not have been pleased. In the early 1800s, the Scottish doctor wrote that nothing in his long career had provided more satisfaction "than my exertions in early life to rescue infants from the cruel tortures of swaddling." The author of an early and earnest child-care guide addressed to mothers—titled, very reasonably, *Advice to Mothers*—Buchan believed that he'd helped eliminate swaddling from the Western world, not unlike how infectious disease specialists today think they've eradicated polio. Buchan would have thought this analogy apt. For him, swaddling the infant like "an Egyptian mummy" caused all manner of alliterative problems: "dwarfishness, deformity, diseases, or death."

So he would have been crushed to observe what happens in hospitals today. As Buchan saw it, the march of progress—the march of liberty even—had freed babies from their straitjackets. But today, several centuries later, more newborns around the world, even in the Western parts of it, are swaddled than are not. Even more babies would be if not for hot, humid climates, which make swaddling inadvisable (rashes, fungal infections, overheating). For infants elsewhere, before freedom there is restraint. Somewhere along the way, the march of liberty got itself turned around.

Swaddling is as old as Western civilization itself. The most famous reference to it is found in the New Testament: in the words of the King James, "Ye shall find the babe wrapped in swaddling clothes, lying in a manger." Like Isaiah's sailing outfit, swaddling clothes in the ancient world were mandatory: every child, Jesus included, was swaddled. Luke left out the salt, though: in the ancient world, before the swaddling bands were applied, newborns were "salted," which amounted to a quick dip in salty water. The salting was thought to harden their delicate skin, the swaddling to cushion their arrival in this strange new land. (Even today in rural areas of Greece and Turkey and much of the Middle East, it is not uncommon for newborns, once at home, to be rubbed with salt before bathing: just as salt makes pork into prosciutto, it is said to "cure" a baby's skin, preserving it for the future.) At the time of the New Testament, salting and swaddling were simply marks of decent parenting. In the book of Ezekiel, a person is put down with this matter-of-fact dismissal: "thou wast not salted at all, nor swaddled at all."

The practice survived more than a millennium of social upheaval. In much of medieval Europe, where snugly bound children spent most of their first year in a cradle, swaddling was as common as it had been in Judea and it was still done with thin linen or wool bands, crisscrossing the infant until taut. It was a complicated, cumbersome process, far more time-consuming than swaddling today, with a single square of cloth. It was also far more restrictive: a well-swaddled infant could scarcely move his head, which was lashed in a stationary position with yet more bands.

The swaddling kept the infants warm, if possibly too warm, and protected from accidents and animals. For a harried housewife, swaddling made a newborn much more manageable: almost anyone, no matter how incompetent or distracted, could look after a swaddled infant. They didn't even have to change the swaddling. Physicians recommended that a swaddled baby be freed and changed every twelve hours or so, but swaddling was so complicated that the advice was rarely followed. Wet diapers, especially thin layers of lousy cloth, which sucked up very little moisture and spread lots of it, likely stayed wet for a long time, out of sight underneath all the swaddling.

In much of sixteenth- and seventeenth-century Europe and America, infants were so tightly wrapped for the first three months that they were more or less unable to move. Today we worry that a baby is psychologically impressionable; the worry back then was that he was *physically* impressionable. The limbs of an infant who wasn't wrapped, it was feared, might "easily and soon bow and bend and take diverse shapes." For a largely rural population, the relevant metaphors were agricultural: the newborn was a seedling and was tended as such. "In the swaddling of it," the midwife Jane Sharp advised in *The Midwives Book: Or the Whole Art of Midwifry Discovered*, "be sure that all parts be bound up in their due place and order gently without any crookedness or rugged foldings; for infants are tender twigs and as you use them, so they will grow straight or crooked." Straight limbs would prepare a baby for a life spent upright; improper swaddling, as a leading French obstetrician wrote in 1668, would mean an infant "would go upon all four as most other Animals do." Conveniently, swaddling *made* infants upright beings: their swaddled form was so rigid that they could be propped up vertically, as if they'd precociously learned to stand.

All this is why swaddling, as the historian Simon Schama observes, has been "taken by historians of childhood as one of the signs of a premodern approach to infancy. Swaddling has been thought to represent the kind of manipulative convenience that betrays parental indifference or self-centeredness rather than a more affectionate and benign attitude towards the child." The sharp decline of the practice, beginning in England and the United

States during the eighteenth century, is often heralded as evidence of a newly progressive, humane approach to child rearing.

But it likely had more to do with a revolution in political philosophy. Swaddling had been popular in part because it restricted an infant's free will—a child left at the mercy of his own appendages might tear out his eyes or ears, it was feared. Suddenly, swaddling was unpopular precisely *because* it restricted the infant: once feared, the freedom of the infant was now prized. John Locke, the prototypical liberal philosopher of child rearing, argued very rationally against swaddling in *Some Thoughts Concerning Education*, but it was the later and somewhat less rational zeal of Rousseau that made the bigger impact. "The infant is bound up in swaddling clothes, the corpse is nailed down in his coffin," he wrote in *Emile*. "All his life long man is imprisoned by our institutions." The child was more free in the womb than he is swaddled, he went on. "Is not such a cruel bondage certain to affect both health and temper?" Rousseau, characteristically, did not know quite where to stop his argument: "We have not yet decided to swaddle our kittens and puppies; are they any the worse for this neglect?"

The flight from a child-care tradition can feel like a toddler's tantrum. It's not that swaddling was a bad idea. It's that it was the *worst* idea! In fact, the opposite of swaddling would be the very *best* idea! Did swaddling keep children too warm? Very well, then: children should be kept *as cold as possible*. "The truth is, a newborn child cannot well be too cool and loose in its dress," the progressive English physician William Cadogan claimed. He added, not reassuringly, that "there are many instances, both ancient and modern, of infants exposed and deserted, that have lived several days." Rousseau himself advocated steadily lowering the temperature of an infant's bath until "at last you bathe them winter and summer in cold, even in ice-cold water." Locke extolled the virtues of leaky shoes.

In the United States and Europe, this new opinion of swaddling—*the worst idea ever*—held for several centuries. Any mothers who still swaddled their children were more or less guilty of criminal endangerment. In 1870, the American etiquette guide *The Bazar Book of Decorum* soberly informed its readers that until swaddling is abandoned, "the health and vigor of whole generations will continue to be sacrificed." By the early 1950s, American mothers interviewed about swaddling reacted with reflexive horror: "An outcry of indignation arose that it would be all-around cruel and unacceptable because it would mean frustrating and restraining the child." (It was puzzling, the interviewer noted, that in so many other countries, "the idea of not giving proper support to their babies, keeping them protected and warm in this way, would be just as abhorrent as the idea of restraint to American mothers.") Swaddling was *foreign*. It was anti-freedom, anti-American: it would erode what made us great.

At the height of the Cold War, the anthropologist Geoffrey Gorer put forward the "Swaddling Hypothesis," which claimed that the foundation of the Russian character was shaped by swaddling. The Russian style of swaddling, especially among the peasantry, was extremely harsh and lengthy, Gorer noted. For the infant, the constriction "is felt to be extremely painful and frustrating and is responded to with intense and destructive rage, which cannot be adequately expressed physically." That rage bleeds into guilt, leaving a psychological imprint on the population: it taught Russians to need authority and to be passive and obedient before it. It was why Russians could not handle freedom and why they needed a strong leader—someone, like their swaddling, who pushes back.

As a national character study, Gorer's hypothesis today seems like a Cold War fever dream, a comic projection of our own agitprop. But at the time it was taken quite seriously. The anthropologist Margaret Mead, by then very famous, defended Gorer, arguing that although swaddling may not be solely responsible for the Russian character, it nonetheless is how Russians "communicate to their infants a feeling that a strong authority is necessary."

Mead was not alone in supporting Gorer. But other academics were more than happy to point out the many problems with the swaddling hypothesis. Among them: if the swaddling hypothesis was correct, then it also had to explain the character of the populations of Europe and the Middle East for much of the previous millennium and beyond. And swaddling, even very tight swaddling, was hardly confined to Russia. In much of the Middle East, not to mention Eastern Europe and Central Asia, the practice persisted. The Western hostility toward swaddling was the exception, not the rule.

In the Cold War West, swaddling was regarded so negatively that the idea there might be more to it than selfish convenience was novel. "Could it be, for example," the authors of a 1965 medical paper on the subject asked, tentatively, "that even nomadic people in ancient times bound their young not only for protection against the cold and ease of handling and transport, but also because this practice resulted in quiescence and apparent comfort?" Indeed, they said, it could: their experiments demonstrated that swaddled infants slept more often and for longer stretches. "It is conceivable," they concluded, "that swaddling may someday return in Western societies as an acceptable general child-care practice during the early weeks and months of life."

And it did: in the last couple of decades, there's been a revival of swaddling "in societies where it was virtually abandoned," as a recent review of the literature noted. The American mothers interviewed in the early 1950s would be shocked to learn that their own granddaughters are swaddling their newborns. In many neonatal wards today, like ours, nurses teach parents how to swaddle. Far from being cruel and unusual punishment, swaddling is now described as a practice that delights babies. *The Happiest Baby on the Block,* the popular baby manual, talks about swaddling in the manner of someone trying to sell you real estate in Florida: it is so enthusiastic you assume you are being conned.

The resurgence of swaddling has inspired a wave of research

into its physiological effects. The studies confirm centuries of anecdotes. For infants under a half year old, swaddling clearly promotes more stable sleep: infants sleep longer; they wake up less; they go back to sleep on their own more. They startle less frequently, at least partly because their startle reflex is physically inhibited—they *can't* startle. Researchers have speculated that the *touch* of swaddling—the feeling of envelopment it engenders—is what makes it so effective. Swaddling may appear unnatural, but our bodies react to it as if it were the most natural thing in the world. Our intuitive sense of what's "natural" is at odds with our deeper physiological responses.

In a way, though, all this randomized research addresses the wrong questions: the real quarrel with swaddling was not that it didn't make life easier for the infant and the parents. It was that it must have pernicious effects afterward—Gorer wasn't the only person who thought that the infantile experience of swaddling reverberated through the rest of life. Gorer's hypothesis was absurd but his logic was banal. Gorer was a Freudian and Freud has seeped into all of us: we believe in the outsized importance of early experience. If a child is bound for much of his early life, it must leave a mark on the psyche. If even merciless swaddling leaves no permanent mark, then what are we all worrying so much about? What *would* leave a mark?

Psychoanalysis aside, swaddling would seem, at the very least, to affect motor development: not being able to move must set a baby back. But nearly every study suggests otherwise. The most detailed study of intensive swaddling is of the Navajo, who traditionally bound their babies to a cradleboard for much of their first year. They were tied to cradleboards a few days after birth and for the next six months, most remained on the cradleboard around two-thirds of the time. The amount of time spent on the cradleboard dropped sharply after the six-month mark, but Navajo babies often stayed on the cradleboard for almost a year, and occasionally even longer. But their motor skills were unaffected.

Why doesn't this binding delay reaching, or crawling, or walking? Why doesn't it harm later motor development?

The answer seems to be that our picture of the cradleboard, and of traditional swaddling in general, is incomplete. By the time an infant is old enough to try to crawl or cruise, he's usually off the cradleboard when he's awake—he may still be swaddled for half the day, but most of that time he's asleep. That's why detailed studies of the Navajo found that cradleboards didn't affect development: the child still got plenty of exercise. Swaddling actually encourages isometric exercise—the sort of workout you get from pushing against something—which means that, counterintuitively, tight confinement may *help* an infant develop his muscles. "In spite of the apparent logic of the premise," as the anthropologist James Chisholm, who studied the Navajo, puts it, there's just not much evidence that swaddling inhibits any sort of motor development. "Indeed, the academic interest in this question of cradleboard effects must now be why it *does not* seem to have any long-term effects." A recent randomized trial in Mongolia supports Chisholm's conclusion: it found that infants swaddled in the traditional Mongolian manner—bound tightly from the neck down for the first few months, then more loosely for another half year or so—experienced no motor delays.

Even cultures that rely or relied on tight swaddling, like the Navajo, never bound the child for the entire day. The idea of the imprisoned baby is more myth than history. When the swaddled baby wants out, he's let out. Other assumptions about swaddling, like other assumptions about other seemingly unjustifiable child-rearing practices, have turned out to be mythical, too. In fact, when you read the many cross-cultural accounts of swaddling, what comes through most clearly is the power *and* the superficiality of differences in child rearing. The power because the differences are so striking: when I see a photograph of a Navajo child bound in a cradleboard, I still feel, despite myself, the urge to spring her. The superficiality because all these accounts, in the end, circle back to the same motivation: to love and care for a child in the best way you know how, and to *make her stop crying please.*

For centuries, there's been a lot of hysteria about *the right way* to take care of babies, and swaddling has been the focus of a lot of this hysteria. Babies *must* be swaddled! Babies must *not* be swaddled! Babies must be swaddled—but not like *that!* While everyone was screaming in italics, the babies themselves seem to have done just fine. Despite their inability to do almost anything on their own, infants are far more flexible than they get credit for: within a few obvious parameters—food, shelter, love—they are astonishingly adaptive. They do well swaddled, very tightly, for long stretches of time; they do well not swaddled at all. And their parents, despite appearances—and there are so many awful appearances—are not monsters: the weirdness of a child-rearing strategy doesn't correlate to its worth. In the history of childhood, no one gets enough credit: not the infants, not the parents.

Please Don't Touch the Baby

There are situations beyond the adaptive power of infants, though. Among the few places that failed to provide babies with the few things they needed was the American hospital and orphanage, circa a century ago: there was no more dangerous place for babies than the very place that was supposed to keep them safe. The Navajo cradleboard may have looked like a primitive torture device, but its effect was entirely benign. The institutions of the new scientific era looked entirely benign, but their effect was that of a primitive torture device.

The scientific era of child rearing had a ready supply of strong, unyielding convictions about how infants must be treated—yet more beliefs about *the right way* to do things. Prominent among these was the conviction that touching a baby was a very, very bad idea. According to the medical thinking of the time, the secret to good health was in limiting your exposure to other people. Since no one was more vulnerable than a newborn, a newborn's exposure to people should be especially limited. The logic of the strategy was inexorable: if a little contact with other people is bad, then no contact at all must be best. This applied not only to babies who were just born, or babies who were sick, but to all babies. And it applied to every sort of contact.

By limiting exposure, physicians were trying to prevent germs from hopping from patient to patient. At this point in medical science, germ theory was so vague that it was more like a theory of ghosts: doctors knew there were these tiny disease-carrying things, but they didn't know *what* they were, exactly, or how they transmitted disease so swiftly and brutally. Orphanages, then called foundling homes, provided particularly brutal evidence of the state of medical knowledge. Infections in foundling homes were so rampant that statistics on survival rates were a cruel contradiction in terms: there weren't any survivors to compile survival rates from. Some institutions apparently had mortality rates of 100 percent: they killed every child they admitted. The agent of death was invisible; the fact of it was omnipresent.

But the problem wasn't just that these infants were catching too many germs. It was also that they weren't catching enough of anything else—the love and attention, the *contact*, that any infant would normally receive. There was a term for the relentless, mysterious decline many babies experienced in hospitals: it was called "hospitalism." The condition was sometimes so bad, the physician Harry Bakwin wrote, that "whether all perished or only most of them depended principally on one factor: the duration of their stay in the institution." As Bakwin noted, "The rapidity with which the symptoms of hospitalism begin to disappear when an afflicted baby is placed in a good home is amazing."

Shortly after the turn of the century, at a time when unsanitary conditions were widely believed to be the biggest threat a baby could face, the pediatrician Henry Chapin made a deeply counterintuitive decision: he began to send newborns *out* of the hospital. He wanted all infants discharged as soon as possible. Chapin had looked at the antiseptic environment of the hospital and seen death. Only a few other doctors were capable of the same leap. These doctors, including Harry Bakwin, realized that hospitalism wasn't caused by filthy conditions or lousy nutrition, or at least not only these things. Its root cause was an absence: the abject deprivation of human contact and attention. At Bellevue Hospital in the 1930s, Bakwin took down the signs in the infant ward reading, "Wash your hands twice before entering this ward." He replaced them with this sign: "Do not enter this nursery without picking up a baby."

This was a radical statement: at the time, many physicians would have read it as an engraved invitation to disease. They believed touching caused death; it didn't prevent it. Outside of Bellevue and a few other hospitals, the war against germs became more and more desperate, as if doctors really were at war with ghosts. Descartes famously argued that a body is a machine, and as the sociologist Anthony Synnott has noted, many physicians at the turn of the twentieth century seemed "determined to put this theory into practice by advocating that children be raised with mechanical precision." If you follow the argument that far, touch is utterly unnecessary. There's no point in touching a machine.

In *The Care and Feeding of Children,* Luther Holt, the premier pediatrician in America at the turn of the century, distilled the turn-of-the-century scientific consensus for nurses and parents. "Are there any valid objections to kissing infants?" Holt asked rhetorically. His answer:

There are many serious objections. Tuberculosis, diphtheria, syphilis, and many other grave diseases may be communicated in this way. The kissing of infants upon the mouth by other children, by nurses, or by people generally, should under no circumstances be permitted. Infants should be kissed, if at all, upon the cheek or forehead, but the less even of this the better.

The medical phobia of touch became a much more general phobia: excessive touching was now seen as having vague psychological consequences. The psychologist John Watson, in his 1928 behaviorist manifesto *Psychological Care of Infant and Child*, strenuously argued against almost any touching at all. "Never hug and kiss them, never let them sit in your lap," he wrote. "If you must, kiss them once on the forehead when they say good night. Shake hands with them in the morning. Give them a pat on the head if they have made an extraordinarily good job of a difficult task."

This is probably Watson's most infamous passage, for very obvious reasons, and it is often quoted. Most versions of it cut off after his jarring advice to just "shake hands" in the morning. But the next sentence is even more remarkable, because it is here that Watson makes what for him is a huge concession: he admits that, in extraordinary circumstances, some sort of physical contact may be necessary. How excruciating he must have found such a compromise! The weakly sentimental parents of today! In any such extraordinary circumstances, this is what Watson allowed parent and child to take comfort in: *a pat on the head.*

These days, touch is what entire philosophies of child rearing are founded on: the baby should not just be touched—she should be touched as much as possible. In Watson's era, touch is what entire philosophies of child rearing were founded against: the baby should never be touched. But neither philosophy is skeptical of the power of touch. For Watson, the problem with touching is that it meant *too much*: to touch the child was to spoil the child, or at least to condition the child to being spoiled. Watson argued that his studies had proven that love could be elicited in an infant "by just one stimulus—*by stroking its skin*. The more sensitive the skin area, the more marked the response." For Watson, touching was dangerous not because it was frivolous or ineffective. It was dangerous because it worked too well: it made children feel love. "Love reactions soon dominate the child. It requires no instinct, no 'intelligence,' no 'reasoning' on the child's part for such responses to grow up." Love was poor preparation for the wider world. (For Watson himself, touching did turn out to be dangerous: he was fired from Johns Hopkins over a very public affair with his graduate student and assistant.)

Watson's attitude was not an aberration: until the 1940s, the conventional wisdom held that touching babies was something to be strenuously avoided—for reasons of character, if not germs. Benjamin Spock, in the first edition of *The Common Sense Book of Baby and Child Care,* marked a new era when he explicitly stated the opposite: "Don't be afraid to love him or respond to his needs. Every baby needs to be smiled at, talked to, played with, fondled—gently and lovingly—just as much as he needs vitamins and calories, and the baby who doesn't get any loving will grow up cold and unresponsive."

But the combination of germ theory, with its fear of touch, and behaviorism, with its fear of touching, cast a long and lonely shadow. In many infant wards and orphanages, babies were "prop-fed"—their bottles propped up so they could suck without having to be held, like gerbils drinking from upside-down water bottles in their cages. The image is heartbreaking.

The Nature of Love

It is also eerily evocative of a picture taken by the psychologist Harry Harlow in a primate lab at the University of Wisconsin in the mid-1950s—not long after infants in neonatal wards were being treated like gerbils. The photograph shows an infant rhesus monkey clinging, with all the strength of its scrawny limbs, to a folded-up towel with a cartoonish face on top, staring impassively down at the monkey. The infant monkey's eyes are bugged out, its expression imploring. The picture is the visual equivalent of nails on a blackboard: your whole body recoils from it.

The photograph captures an iconic experiment, a study that overturned an entire generation of thinking on what mothering means. It was a brutal experiment. Today, depending on who you ask, Harlow is either acclaimed (as a pioneering researcher) or despised (as a merciless torture artist).

But we are all, unavoidably, his descendants: it was Harlow, a half century ago, who laid the empirical foundation for what seems self-evident today—he linked touch to love. He did so in a cruel, wrenching, animal rights movement–inspiring way, especially in his long-term isolation studies. The irony of this idea being verified in this way is not inconsiderable.

In his most emblematic experiment, Harlow, working with his graduate student William Mason, placed infant rhesus monkeys in a cage with a couple of what he called "maternal" figures: a soft, terrycloth-covered "mother" and a wire "mother" with a nipple that dispensed milk. The monkeys had a clear preference: they only left the soft comfort of the terrycloth mother to get food. Their relationship to the wire mother was strictly utilitarian: once they'd had enough milk, they hopped off. They needed what Harlow called "contact comfort." And when Harlow took the terrycloth mother away, they went berserk.

In a paper modestly titled "The Nature of Love," Harlow argued that his experiments proved that touch, or "contact comfort," was what made a mother a mother. Contrary to Freudian theory, the monkeys attributed no maternal qualities to the mother with milk. They attributed maternal qualities to the mother *without* milk.

Freud wasn't just wrong, Harlow concluded. Freud was really, *really* wrong: "Indeed, the disparity is so great as to suggest that the primary function of nursing as an affectional variable is that of insuring frequent and intimate body contact of the infant with the mother. Certainly, man cannot live by milk alone."

To make a preposterous, unverifiable claim: in our modern era, this is the moment when touch is made synonymous with love. It's odd to think that there was a moment *before* this moment, when touch wasn't synonymous with love. Many parents obviously already knew what Harlow established: an academic discipline is often the last to know. But the astonishing success of John Watson and other child-care gurus suggests that deeply counterintuitive ideas about child rearing had a ready audience. Ever since the dawn of scientific child care, parents had wanted to know the new *right way* to raise children. This wasn't blind faith in science so much as a logical extrapolation from recent history. In a very brief period of time, science and technology had changed everyday life in innumerable radical ways. Why wouldn't parenting be changed, too?

Harlow is a complicated figure in the history of psychology not just because of the brutality of his experiments. He's complicated because his insights don't seem like insights. He demonstrated that the mother, not the milk, was what mattered. And he established the need for what he, to the irritation of many other psychologists, insisted on calling *love*.

Harlow's contrary idea salvaged another contrary idea: attachment theory. Outlined by the psychiatrist John Bowlby in the early 1950s, attachment theory is best known today for its cousin, attachment parenting, a method of child rearing with passionate devotees and detractors. It's what the pediatrician Robert Sears is always on about. But the concept of attachment was born in a report commissioned in 1949 by the World Health Organization on the mental health of homeless children in postwar Europe.

It's an unlikely genesis for an idea that would change how many people thought about child development. But Bowlby's WHO report was not dry, or bureaucratic, or hedged. It was written in fire. Modern society, he argued, overlooked the most important factor in how children develop: a mother's love. The deep connection to a mother, or at least a maternal figure, was essential for a young child's well-being, just "as are vitamins and proteins for physical health," and children deprived of it would be severely, and sometimes irreparably, damaged. Furthermore, Bowlby went on, as parents themselves, they would inflict that same damage on their own children. It was a hugely influential report, and a couple of years later, republished as *Child Care and the Growth of Love,* it became what must be the only WHO report to end up an international bestseller.

In the decades that followed, Bowlby would elaborate and refine attachment theory, and although it has many complexities, it was never complicated. Stripped down to its bare essentials, it proposes, very simply, that an infant needs to be with someone who provides love and is always there, and that this first relationship is the basis for emotional well-being. You get security as an adult (and security in this definition amounts to pretty much everything) by being securely attached as a child.

Who that person should be, and if there should only be one person, would become fraught topics. Bowlby and attachment theory eventually inspired a lot of feminist venom, in part for some troglodyte statements Bowlby made about where mothers belong. But Bowlby was always less invested in what mothers should do than he was in what infants need: the emotional demands they have that can't be waved away. These days, the core of Bowlby's argument seems like common sense. It's strange to even call it an argument: *children need love.* Who'd take the other side?

But the postwar world was deeply unconvinced by it. Attachment theory was sentimental, loosey-goosey, fantastical, not supported by evidence. It was not until Harlow's experiments that many realized perhaps Bowlby was on to something. Maybe touch—maybe having someone to cling to, to hold on to—mattered.

12

A Blessed Break from Emotion and a Sudden Detour into Physiology, Carnival Midways, and Kangaroos

We tend to think of organs as discrete and internal: the kidneys, the gallbladder. But the skin is an organ, too—it is the Epic Western of the genre. Flayed, an adult's skin would weigh in around nine pounds, depending on the specimen. It would stretch over eighteen square feet. It is huge; it is bewilderingly complex; and it is implicated in almost everything we do. "If you think about it," the writer Diane Ackerman says, "no other part of us makes contact with something not us but the skin." And yet, despite this Zelig-like omnipresence, skin is the organ we are least likely to notice—except when there's something wrong.

With premature infants, it is often very clear that something is wrong.

Mature skin relies on the stratum corneum, the top of the top of the epidermis, a film as thin as twenty micrometers, or twenty-millionths of a meter. Within those micrometers are as many as twenty separate layers, which form the first and best barrier against toxins, bacteria, and viruses. They perform myriad essential functions, like the regulation of heat and water loss.

But the stratum corneum only develops during the third trimester, which means that infants born at the limit of viability, today around twenty-four weeks, may have none at all. Infants born around thirty weeks have a gossamer coat of it. Premature infants sometimes look as if they are on fire because their blood vessels are almost at the surface, turning their skin a shimmering, angry red that screams pain. The skin is so thin that a few bits of tape can do excruciating damage, a major problem since premature infants have innumerable medical monitors affixed to their bodies.

The stratum corneum is meant to be somewhat permeable. It stands guard over what goes in and out of the body. But in premature infants, the layer is not yet up to the job. It lets out too much heat and water; it lets in toxins. The most fragile premature infants sometimes lose fifteen times more water through their skin than full-term infants do. With water goes weight: in worst-case scenarios, 30 percent of total body weight can be lost in the span of a single day.

The regulation of temperature and fluids is vital: almost nothing is possible without it. But the age of viability is now so early that the skin of seriously premature babies is challenged by even these basic functions. It is strange to think that the immaturity of skin could be a major barrier to survival: it seems like there must be something much bigger and more complicated that goes wrong first—some *other* organ. Premature medical care has made staggering advances—surgery on congenital heart defects, for starters—only to run up against the bare fact of skin.

It would not seem especially likely, in short, that touch would be a salve to babies born prematurely. Almost no one thought it likely. But the benefits of touching turn out to be felt most profoundly by preemies: those most vulnerable to touch are also those who demonstrate how powerful it can be.

The Invention of the Premature Baby

A century ago, there was no preemie ward. The first ones opened in the early 1900s—and closed shortly thereafter. They were no use. Any baby born seriously premature was beyond the powers of medicine. There was also the question of whether babies born so early *should* be saved, even if they could be: perhaps, many thought, they weren't meant to live. They were called, in language borrowed from social Darwinism, unfit to survive.

The invention that helped change such thinking was hardly even an invention: it was a tweak of a device in the baby chick display at the Paris Zoo in 1878. The French obstetrician Etienne Tarnier, who'd seen it on a visit to the zoo, asked the zoo's director to build him a version of the device for human infants. Voila: the preemie incubator.

Perhaps understandably, given this pedigree, the incubator wasn't received as anything revolutionary. In American hospitals, it was either ignored or disparaged. Pediatricians were deeply suspicious of incubators: the invention ran contrary to their belief that fresh air and clean food were all a baby, even a premature baby, really needed. (Harry Chapin, who was so quick to see that many newborns were better off outside of the hospital, argued that "incubators should be abandoned entirely.") Many hospitals rejected incubators outright; they weren't widely used until the 1940s, over a half century after their invention. In fact, many Americans first saw the incubator in a place about as far from the hospital as can be imagined: the carnival midway.

The carnival displays were the peculiar genius of the German doctor Martin Couney. The fact that premature infants *could* survive was so surprising to people, even fellow physicians, that Couney, trying to promote his incubators—his *Kinderbrutanstalt*, or child hatchery—took them out on tour. Premature infants had such low rates of survival that many parents and physicians were happy to hand over the babies to Couney, at least on a short-term basis. On display at fairs and carnivals, they were a neo-natal freak show that put preemies—*living babies*, the advertisements proclaimed—on the same level as the bearded lady and the sword swallowers. At the Chicago World's Fair in 1933, the *living babies* exhibit, populated by infants borrowed from a local hospital, sold the second-most tickets at the fair. The most: a burlesque dancer with ostrich feather fans.

Couney's success at caring for premature infants created its own problem. He had no trouble getting the preemies; his trouble was in returning them. After weeks or months without contact with their infants, mothers felt that they weren't really *theirs* any-more. They didn't want them back. This was the paradox posed by the incubator from its invention: to make the infant fit for the world, the infant had to be removed from the world. To survive in the short term, the infant was cut off from the people his survival would depend on in the long term. The irony was apparent from the beginning of premature medical care. In the first incubator experiments ever conducted, the French obstetrician Pierre Bu-din, the protégé of Tarnier, explicitly took note of it: "Mothers separated from their infants soon lose all interest in those whom they are unable to nurse or cherish."

Decades after the no-touching policies of neonatal wards were abolished, they persisted in a place where they seemed to make a lot of sense: the preemie ward. Before the 1970s, all premature infants were kept in abject isolation. "When I was a pediatric resi-dent" in the 1960s, the neonatologist Alistair Philip has observed, "I was barely allowed to enter the premature nursery and parents were allowed to view their infants only through a glass partition." The only touching allowed was for medical purposes, which meant that the only touching preemies ever experienced was strictly clinical and often painful. A study of a NICU concluded, "Approaching an infant for the sole purpose of providing social stimulation was a rare event."

In the early 1970s, a few doctors made the case that incubators, by cutting parents off from their child, hurt the chances of recovery even as they made that recovery possible. They argued that premature care had to be rethought: that the mother of a premature infant has the same emotional needs as the mother of a term infant—she needs to feel that her infant is hers. There may not be empirical proof of this, an editorial in the *British Medical Journal* acknowledged. But we should not need it: "No one has proved that it is desirable for the mother or the premature baby that this close contact should be established immediately after birth or later during the period in hospital or that absence of contact does any harm. One cannot prove everything, and not everything is worth trying to prove."

An experiment in which mothers were taught proper hospital hygiene and allowed to handle their premature infants at any time proved that it could not hurt: the touching by itself did not increase rates of infection. Moreover, the mothers who'd touched their infants were more confident, more committed, more engaged than those who had not touched theirs.

Touching was not the specter of death, after all. It was the spark of life. As the *British Medical Journal* editorial put it, with unexpected grace, "It would be natural and normal for a mother of a premature baby to be helped to feel close to him, to feel that the baby in the box is hers, to maintain contact with him in hospital from the time of delivery to the time when she leaves for home where she will have to do everything for him—and to feel the great satisfaction that she has helped to save her child from the valley of death."

Kanga and Roo: The Maternal Pouch

This sentiment was more true than the *British Medical Journal* knew. It wasn't just that "maintaining contact" didn't harm. It could help; it could be a cure. But it took a desperate doctor, in a desperate hospital, caring for desperately ill newborns to discover this.

The San Juan de Dios hospital in Bogota in the 1970s was the typical overburdened hospital in a poor country: too little money, too few staff, too many barely alive newborns. The impoverished women the hospital served often received no medical care during their pregnancies; many gave birth prematurely. There were more babies who needed incubators than there were incubators. Newborns were doubled up; sometimes three were squeezed into a single incubator. Infection was rampant and heat was nonexistent. Many infants, before they died, were simply abandoned. The doctors of San Juan de Dios did the best they could. It wasn't enough: the mortality rate for premature infants was 70 percent.

None of this was unusual. The developed world has the resources that prematurity requires: incubators, ventilators, expertise. What it does not have are the premature babies. Some 95 percent of newborns with a low birth weight are born in the developing world. Very few who need intensive care have access to it. The consequences are predictable: across the world, prematurity is the single largest cause of infant mortality.

The San Juan de Dios hospital was destined to be a very poor hospital caring for very high-risk infants, almost all of whom it would lose. But the physician in the neonatal intensive care ward, Edgar Rey Sanabria, was desperate enough to try something truly strange: he instructed the mothers of premature infants who could at least breathe outside the incubator to hold their babies against their bare chests, underneath their clothes, every hour of the day and night. It would at least keep the babies from being too cold, Rey thought, and it might make breast-feeding easier. If the infants stabilized, Rey had the mothers continue the procedure at home. He didn't have much to lose. Was his ward's mortality rate going to go *higher* than 70 percent?

The new regimen was successful beyond imagining. The infants who were held had far fewer infections. They nursed more; once home, they were hospitalized less. And far more survived: the mortality rate dropped to 30 percent.

Now known as kangaroo care, or skin-to-skin care, Rey's last-ditch idea has been implemented in countries across the world, including places where doctors have plenty of more complicated options. It is the most affordable, most effective way to reduce prematurity-related deaths and illnesses in developing countries, where there is far too little equipment but plenty of warm skin. Flesh is what's always on hand.

In the thirty years since Rey told a mother to nestle her tiny infant between her breasts, skin-to-skin care for preemies has run the gauntlet of modern scientific evaluation: randomized trials, peer review, replications. It's held up and physicians now know a lot about the technique that seemed highly questionable at first. An example: being skin-to-skin develops a baby's basic self-regulatory capacities, the sort of skill that is usually fine-tuned in the womb. It's a crucial skill. The better a baby is able to regulate his responses to this strange new world, the less of his precious energy is wasted: he puts on more weight; his cognitive development speeds up.

But there remains an abiding strangeness to the idea of even *evaluating* kangaroo care as a new medical treatment. It's just a mother pressing her child against her bare skin. At its core, the act isn't novel. It's more like primeval.

Such a premodern idea could only have originated in a place like Bogota. In a modern hospital, it would have violated too many assumptions about what premature infants can and can't handle. Kangaroo care is not a little sweet antiseptic stroking. In its "continuous" version, an infant is pressed against the chest, head between the breasts, appendages splayed, continuously; he's held snug in place by a wrap. The name "kangaroo care" is meant as an exact description: Rey was inspired by marsupials. In the history of modern medicine, when people are inspired by the child-rearing behavior of other species, things generally turn out very badly. Kangaroo care is the rare exception.

It is, fundamentally, an attempt to reverse time: to put a baby who came out of the womb too soon *back in the womb*. Curiously, physicians sometimes miss this point. A recent review of the technique describes the role of the parent as "somewhat like human incubators, providing physiological homeostasis, appropriate stimulation, and the main source of nutrition." That's backward, of course: the whole point of an incubator is that it functions somewhat like a mother. To say that a mother functions like an incubator is to credit the incubator with far too much originality.

Being skin-to-skin has a nearly instantaneous effect on almost all newborns, from the very premature to the nearly term: they go out like a light. The few who manage to stay awake slip into a quiet, subdued state. They stop crying; they stop protesting. In fact, extended skin-to-skin contact seems to mold easier babies: many studies show that infants who were given kangaroo care are calmer and less agitated, as much as six months later.

Kangaroo care does not give premature infants in developed countries better odds for survival. But it does offer something that technology cannot: intimacy. The first medical evaluation of Rey's methods, published in *The Lancet*, recognized this virtue immediately. Skin-to-skin care may, the article noted, "help us to heal some of the psychological problems incurred by modern neonatal care."

It was a prescient assessment: later studies have shown premature babies who are held against their mothers for an hour each day for a couple of weeks are *psychologically* better off in almost every way. Their mothers are more sensitive and responsive and warm, even while in the hospital; they are less likely to be depressed later on. The infants are more socially alert. Skin-to-skin care of premature babies reverses the negative feedback loop of prematurity, the way out-of-it infants beget out-of-it parents. The feedback now runs the other way: being closer to their premature infant makes parents more sensitive and aware, which in turn makes the infant more aware, which in turn . . .

Remarkably, babies given kangaroo care seem to sense when their time in the womb would have been up: at right about their due date—just before the forty-week mark—they strain against the claustrophobia of the makeshift pouch. They get antsy; they want out. Even for infants born at term, though, kangaroo care, if used for short periods of time, leaves an imprint: after skin-to-skin contact of only an hour, infants sleep better, cry less, fuss less. Their movements are smoother, less abrupt; their body temperature creeps up. Extended skin contact after birth forms a sort of stress-reducing shield against the total chaos of life on the other side of the uterus. A baby who's tucked into his mother is in a quieter, less hectic world.

The evidence for kangaroo care continues to accumulate: a recent meta-analysis—a sophisticated review of all relevant studies—concluded that the technique "substantially reduces neonatal mortality amongst preterm babies" and is "highly effective" at reducing deaths from infection. Holding a baby born too soon against your chest, the simplest of all medical interventions, has the power to "substantially reduce" the one million deaths attributable every year to premature birth.

But even where it would make a real difference, kangaroo care is still too rare: very few premature infants in developing countries receive it. These infants are unfortunate twice over: not only are they not lucky enough to be born somewhere with a modern neonatal intensive care ward, they aren't even somewhere with the no-cost substitute for that neonatal ward.

Stressed Since Birth: Infant Massage

In the early 1970s, an American woman named Vimala McClure, working at an orphanage in India, was inspired by the local traditions of infant massage. When she returned to the United States, she published the first how-to guide on the subject: *Infant Massage: A Handbook for Loving Parents*. Infant massage was almost unknown in the United States at the time; there was no obvious readership for a guide to it. But the passive-aggressive subtitle must have worked. Today, over three decades later, there are so many guides to infant massage that one new entry, *Hands on Baby Massage,* differentiates itself by including a onesie "preprinted front and back with a colorful design that shows just where to massage for specific results."

Like swaddling, once seen as deeply *other*, infant massage is suddenly ubiquitous. It's a ritual with ancient roots in China and South Asia, places where, until recently, not massaging your baby would have been simply unthinkable. Its roots here are still shallow but I kept tripping over them: our hospital has classes in it; our pediatrician has flyers about it; the baby blogs blog about it.

Ubiquity works. Before Isaiah was born, I was convinced that taking an infant massage class was a Smart Parenting Decision. We'd ignored the hospital's HypnoBirthing® series, since any childbirth class with a registered trademark seemed like a bad idea. We weren't tempted by the baby sign language classes. We didn't take any of the childbirth classes, or even the seemingly indispensible CPR class. But I was enthusiastic about the massage class.

Until I realized it cost several hundred dollars. After which we decided that our child would have to get used to our crude, clumsy hands.

Isaiah was born in early December. It was mild and oddly bright on the day we brought him home and I remember setting his car seat on the sidewalk and feeling blinded by the light, the way characters are in excessively foreshadowed Hollywood movies. Then winter hit. We celebrated winter by taking him for a walk in his first snowstorm, which he celebrated by getting his first cold. After that we decided to stop celebrating so much. His first few months were a tangle of layered diapers and onesies and swaddling blankets, underneath all of which was, at least in theory, him. But we rarely found him: like peasants in the old world, we were too worried about the cold to leave him without his clothes for long. Before you have a baby, your image of new parenthood is of spending long stretches of time pressed against your child's new skin. This is the mother-baby—and at least occasionally, the father-baby—idyll that we dream of before birth. But it turns out that, at least for us, it was oddly difficult to spend time up against our baby's skin unless we explicitly made time to do so. The main regret I have about Isaiah's early months is that I wasn't skin-to-skin with his naked body enough. There were too many things—the cold, the poop, the itching desire to do non-infant-related work—that got in the way.

Later on I realized that the real worth of infant massage classes isn't the method. It's the excuse. In the same way that daytime talking heads advocate *making time for family dinner,* the point of infant massage is that it requires you to make time for infant massage. Infant massage classes are what give parents permission to spend time lovingly squeezing every fold of their newborn; we somehow weren't able to give ourselves permission. It's strange, in retrospect. You wouldn't think that you need a good reason to spend time lolling with your own naked baby. You would think that the baby *was* the reason.

There are plenty of ways in which infant massage benefits full-term infants: they calm down faster, they sleep better. And in a seminal study, premature infants who were massaged grew nearly 50 percent faster than those who were not and they were discharged from the hospital five days earlier.

Despite such dramatic findings, massage in premature infant wards remains a novelty: human touch is still seen as alternative medicine. According to the psychologist Tiffany Field, who conducted the original study on massaging preemies, neither the nurses nor the doctors can make sense of it. The nurses, she says, think "the babies should be surrounded by a blanket and as little time as you touch them, the better off the baby's going to be." And until there's a physiological explanation for the effects of massage, most neonatologists will be skeptical of it, she says—even though the effects are apparent without any explanation at all.

In a sense, this is the same argument we have heard before. Physicians have repeatedly struggled with whether touch—part love, part germ—is more risk than benefit or more benefit than risk. They have yielded in favor of touch nearly every time. But they're still unconvinced.

"They didn't expect touch to have this power then," says Field, recalling how doctors reacted to her original study, "and they still don't."

13

In Which Touch Gets Perhaps a Little Too Much Power

Here's that Saul Schanberg quote again from above—way above:

> If touch didn't feel good, there'd be no species, parenthood, or survival. A mother wouldn't touch her baby in the right way unless the mother felt pleasure doing it. If we didn't like the feel of touching and patting one another, we wouldn't have had sex. Those animals who did more touching instinctively produced offspring which survived, and their genes were passed on and the tendency to touch became even stronger. We forget that touch is not only basic to our species, but the key to it.

No species! No parenthood! No survival!

It's not that Schanberg is wrong in thinking this, not exactly, although there are a lot of other things that might meet these criteria, too. (Play along at home: Language! Bipedalism!) It's that thinking clearly about how much touch matters is hard to do when people are telling you, *Lord of the Rings*–style, that *it is the ring*.

Not surprisingly, given this rhetoric, touch is where a lot of our anxieties about parenthood take root. And naturally enough:

if touch is where meaning is found in infancy, what does that say about when we're *not* in touch? And if touch is so important, what does that mean for parents? Does it mean that they have to be glued against their newborn? Does a good parent have to drop everything that isn't the baby?

It's not an accident, in other words, that some prescriptions for parenting are founded on touch. We're already primed to think this is true. Some child-rearing books present touch as the solution to practically *any* problem. Proximity itself, as the critic Rebecca Kukla says, has become fetishized: "There is a suggestion that mothering, in all its complexity, can be effectively reduced to nearness, and that the value of mothering is somehow contained in this contiguity of bodies." Closeness is all.

No one today would argue touch isn't crucial to a newborn. There are fundamental socioemotional needs shared by every infant, as Bowlby proposed. Sarah Hrdy has reduced those needs, in a single lovely sentence, about as far as they can be reduced: "Human infants have a nearly insatiable desire to be held and to bask in the sense that they are loved." *To bask in the sense that they are loved.* Ever since Harlow, it has been excruciatingly clear that touch in infancy matters.

If anything, we are now too willing to believe that it matters. Touch was once demonized. It's now deified.

But scratch the surface of some of the claims made for it and your finger sometimes slips right through. This begins even before birth. Touch is said to be the mother sense, the first to develop in the womb, the most refined at birth. This is what everyone—the books, the experts, the blogs—tells you, at least. But you can only follow the trail of touch in utero so far before you get lost. Sensory receptors develop around the mouth as early as seven weeks and then spread across the body: to the rest of the face, the hands and feet, the arms and legs. Do these sensory receptors form a *sense*? Does this fetus experience the sense of touch in the same way you and I do? Seven weeks is extremely early; the fetal neocortex only begins to grow at eight weeks. Very early fetuses have been shown to react to touch—to flinch in response to it—but these reactions occurred outside of the womb, on what medicine gently calls "exteriorized fetuses." Would a significantly preterm fetus in the womb respond the same way? The womb is awash with various biochemicals that may inhibit sensation. The neural systems that transmit sensation are still in development. Babies in the womb clearly have sensory experiences; they are acquainted with touch. But their version of touch may not be the same as ours. It may take a couple of trimesters for a fetus to *feel* in the way we do: a twinge, a tickle, a rub.

But even then the picture is murky. The way a fetus shies away from a probe: is that an unthinking reflex or some sort of deeper awareness? The question of fetal pain, for instance, is hugely fraught, for all the obvious reasons. Doctors now agree that the newborn is a highly sensitive being, able to sense pain. It'd be surprising if the act of being born flipped the switch on that sense; it must be present at some point beforehand. But when? We don't know, not with any confidence.

And since no one knows, why not gussy things up a bit? The fetus is often described as if she were sloshing about inside an amniotic kiddy pool, a bubble of sensation. Outside of the womb, where you'd think things would be easier to see, the picture isn't always clearer.

Bonded at Birth. Or at Some Other Time Altogether.

It is impossible to talk about touch in infancy, to talk about skin-to-skin contact, without talking about the bonding craze of the 1970s: the idea that there is a sensitive period, just after birth, during which a mother and child have a unique opportunity to form a bond and that the strength or weakness of their bond will have lifelong consequences. The postnatal bonding period was a radical idea and an immediate sensation. It changed birthing practices. It changed how new mothers saw their babies, or at least how they thought they should see them. It was touted as very, very important by nearly every expert professing any sort of expertise. And almost all the claims for it have fallen apart.

The too-easy confusion of *skin-to-skin contact* with *bonding* is emblematic of how touch in infancy is talked about: the benefits that are real and the benefits that are phantom are all jumbled together. The long half-life of the idea of bonding—the idea that a child could be forever won or lost in the day after birth—is a case study in how such messes get made.

To understand the climate in which the concept of bonding was credible, you have to skip backward a quarter-century from it, back to when John Bowlby first proposed attachment theory. Bowlby's ideas were shaped by the attitudes of a very specific time and place: the early twentieth-century Anglo-American conviction that children should be raised strictly, without excessive emotion, and that any well-developed child should take swift flight into independence. Bowlby's childhood was prototypical: he'd had a nanny until he was sent off to boarding school, at the advanced age of eight. He despised all of it. The sorrow for his own lost childhood—and the desire to salvage for other children what he hadn't had—lies just beneath all his writings.

The cold convictions of the era are apparent in the visiting policies hospitals had for child patients. At the time of Bowlby's WHO report, a child hospitalized for a week would likely see his parents for no more than an hour—not an hour a day, an hour a week. Many hospitals refused to let parents visit at all. Parents were useless interlopers. What good could come of having them around? At best, they were annoying; at worst, they might spread disease. To many children, the policies were devastating. Robert Karen, in his book on John Bowlby and attachment theory, describes the annual Christmas visits of the BBC to hospitals in the early 1950s: "When the microphone was brought to older children, they happily consented to wish their families a Merry Christmas over the air. But toddlers stared silently and then burst into tears. This happened so consistently that soon the little ones were omitted from this part of the program."

The capacity of physicians to justify these rules, even in the face of obvious distress, was astonishing. James Robertson, a social worker who worked for Bowlby, collected wrenching evidence of the impact on children, but his work was rebuffed by nearly everyone who heard him. "What is wrong with emotional upset?" a pediatrician demanded after Robertson presented his findings. "This year we are celebrating the centenary of the birth of Wordsworth, the great Lakeland poet. He suffered from emotional upset, yet look at the poems he produced."

Against arguments like this, Robertson realized that words alone would not be enough. He needed to *show* what was happening. With Bowlby's help, Robertson randomly selected a child in the hospital—Laura, just shy of two and a half, admitted for surgery—and filmed her for the same forty-minute period, twice daily, during her eight days in the hospital. The methodology had to be unimpeachable; Robertson knew he'd be accused of cherry-picking footage, of only filming the patient when she was upset. The documentary, *Laura*, released in 1952, sparked a firestorm: Laura was a remarkably self-possessed child, but her quiet distress at being without her parents, who were allowed brief visits every other day, was almost unbearable to watch. Doctors who saw the film called for it to be banned. Its integrity was attacked; Robertson was slandered. But slowly, a few English doctors and nurses began to admit that the film was accurate. And they began to look at hospitalized children in a new way. "I was angry," a London pediatrician later described his reaction, "but after the film I really heard children crying for the first time."

This acceptance was halting, though, and it stopped at the Atlantic. When Robertson screened the film to physicians in the United States, he was told that it simply wasn't relevant there.

American children weren't spoiled like English children, the doctors said. They tolerated the separations quite well. In England, change came but not fast enough for many, including Robertson. A decade after *Laura*, Robertson advised parents to practice civil disobedience. "If they decided to stay quietly by the cot of their sick child they could not be evicted," he said.

It was against this backdrop—the refusal of many hospitals, in the very recent past, to admit that young children needed their parents—that the bonding debate played out. In the early 1970s, when the first studies on bonding were published, most mothers were separated from their babies shortly after birth—the newborns were sent off to the hospital nursery, where they stayed until they went home: the nursery was prized as a sterile, safe environment, free from infection and the clumsiness of new mothers. It was more "scientific," too—newborns were fed formula, which could be measured and tabulated, unlike breast milk, which was stubbornly imprecise. The physicians John Kennell and Marshall Klaus, the authors of the "sensitive period" theory of bonding, argued that this policy was backward—that the absence of the mother might cause more serious problems than the regulations were there to prevent. In their 1976 book *Maternal-Infant Bonding*, Klaus and Kennell wrote that shortly after the birth there is "a unique period during which events may have lasting effects on the family." This is when bonding takes place, they said, and it is so important that its success or failure echoes through the entire life of the child.

From the very beginning, the evidence was iffy. In their initial study, Klaus and Kennell divided first-time mothers into different groups: the experimental group—the ones who would be bonding—were assigned an hour of skin-to-skin contact after delivery and five hours on each of the next three days. A month later, Klaus and Kennell assessed the bonding through observations and an interview. The mothers were asked, for example, if they had gone out without the child and how they felt about it. A mother who said she'd gone out and felt fine about it received a score of zero—she was deemed to have not bonded. A mother who hadn't gone out, or who had but worried the entire time, received a three—she *had* bonded.

There are a lot of assumptions behind this experimental design, needless to say, and what you conclude from it depends entirely on those assumptions. As Diane Eyer, a caustic critic of bonding, notes, "There is the question of the extent to which many of the dependent variables—letting the baby cry it out, not going out without thinking about the baby—are actually valid measures of caretaking. A woman who cannot leave her baby might be overanxious or might not have anyone reliable to leave the baby with. A woman who is able to forget about the baby when she goes out might have a trusted babysitter or might be self-assured and highly competent."

Klaus and Kennell had been inspired by evidence for a "sensitive period" in other species. Oddly, though, these were highly dissimilar species: instead of other primates, sheep. (Also elk and goats.) The postpartum behavior of sheep was so compelling—a mother sheep recognizes her lambs by the exact smell they have after birth—that everyone seemed to suspend their critical judgment about what that meant for humans. As go mother sheep, the logic went, so go mother humans.

This logic turned out to be less than scientifically satisfying, but at the time it was persuasive. Many child-care experts were perfectly willing to follow it. Here's the well-known British doctor Hugh Jolly, counseling mothers about the dangers of being apart from your child after birth: "Animal studies of the effects of short periods of separation of mother and offspring have shown disastrous consequences—rejection and even killing of the baby." It was not noted if Jolly, then in his early sixties, had ever in his long career seen a human mother reject or kill her own baby after being briefly separated from him. But Jolly was willing to override his own experience in favor of extrapolations from sheep.

And he was typical: both Benjamin Spock and T. Berry Brazelton made similar statements, before tiptoeing away years later. (John Bowlby, to his credit, was far more skeptical.) A headline in *New Scientist* captured the tenor of the time: "The One-Day-Old Deprived Child."

Meanwhile, parents who hadn't been able to hold their baby after birth panicked, afraid that they'd missed their only chance at forging a lasting bond. "We're all getting along just fine," said a father at the time about his toddler daughter, born via cesarean, "considering we didn't have the chance to get bonded." Even the parents who were able to hold their baby must have panicked about whether they were doing this very crucial bonding thing right. Soon the fervor over bonding was at such a pitch that Klaus and Kennell had to talk parents and other doctors down: even if mothers hadn't been in close contact with their child after birth, they'd still be able to bond. Bonding, they insisted, wasn't an "epoxy" process; it took time. But it was too late: people had already bonded to the original idea of bonding.

When the tide came back in, it was devastating: almost no claims for bonding were left standing. Except that bonding did matter for mothers at risk of walking away from their child. That makes sense, of course: it would be harder to abandon someone you've met than someone you haven't. But the vast majority of mothers *weren't* at risk of abandoning their babies, and for them, post-birth contact had zero effect on whether they and their infants were judged to be "attached" a year later.

Bonding is now remembered as a dizzy time, when everyone got a little carried away. The zoologist Peter Klopfer, whose studies on imprinting helped inspire a lot of bonding research, has written a half-comic apologia for his role in it all. "Was I ever heard extrapolating human maternal behavior from that of goats or lemurs? I was? Well, then, this is a mea culpa."

Mea culpas were necessary. But in a paradoxical way, the whole misguided idea of bonding might have been necessary, too: it was a reaction to hospital regulations every bit as absurd as it was. And bonding won: by the time it turned out that Klaus and Kennell were shooting blanks, the hospitals had already surrendered. For the first time, many hospitals allowed mothers and newborns to stay together after birth. It was once common practice for doctors, who always knew best, to knock out a laboring woman until the baby arrived; now they stopped discouraging women who wanted "awake and aware" births. Before Klaus and Kennell's work, the authority of doctors over a new mother was rarely challenged. A decade later, in the early 1980s, *Our Bodies, Ourselves* pointedly advised pregnant women that "there is no medical reason to separate healthy mothers and babies after birth." Today we take this idea for granted; we *assume* mothers and babies are together after birth. But many mothers and babies were denied this experience, and if not for Klaus and Kennell, even more would have been.

The legacy of bonding is not hard to trace today. On the Dr. Sears Web site, the popular attachment theory–oriented physician admits that the notion of bonding as "an absolute critical period or a now-or-never relationship is not true." But a reader could easily get the sense that Sears still sort of thinks that it *is* true. In the most recent edition of *The Portable Pediatrician*, Sears and his wife Martha, a nurse, write, "Bonding in the first few hours creates physical and hormonal changes in both parents and baby that set the stage for a lifelong healthy relationship."

Is a baby born prematurely or via cesarean "permanently affected" by the separation after birth, Dr. Sears asks? "Catch-up bonding is certainly possible, especially in the resilient human species." This is not reassuring: when a doctor takes shelter behind the word "resilient," things are looking dark. Missing the chance to bond at birth, Sears adds, makes it all the more critical that you follow his advice: "the attachment style of parenting can compensate for the loss of this early opportunity."

But it isn't just Sears who continues to imply that bonding is crucial. A recent academic review of the research on touch and development states, without qualification, that immediate contact after birth is "critical for the formation of the mother's 'bond' to her infant and initiates a cascade of neurological, hormonal, behavioral, and cognitive changes in the mother that follow childbirth and are required for the onset of mothering." When babies miss this "sensitive period," the missed opportunity "has a life-long impact on the development of regulatory functions in the cognitive and social-emotional domains."

It is hard to exaggerate just how confused and misguided this is. But it is bold in its wrongness: you would never suspect it.

You can see the origins of this mishmash in the book that sparked the modern study of touch. *Touching: The Human Significance of the Skin,* written by the anthropologist Ashley Montagu, was published some forty years ago but it is still in print and still widely read, even venerated. Montagu argued that touch is the root of our humanity, back when this seemed like a contentious claim. And Montagu didn't just argue it; he was evangelical about it.

It was a claim that cut against the grain of history. Since the ancient Greeks, touch has been regarded as a lesser sense. The contemporary reverence of touch is an anomaly. By the standards of just about everyone since Aristotle, valuing such matters as the contact between mother and child reflects somewhat peculiar priorities.

It was Aristotle who ranked sight as the highest sense, the supreme sense, and sent touch to languish with the lower senses—those that dwell within the body rather than rising above it. Aristotle's rankings were left intact by almost every philosopher who considered the senses thereafter. Immanuel Kant was typical: he argued that the immediacy of touch—its directness—rendered it primitive and unreliable. Touch was irrational. More removed senses like sight were the source of true knowledge.

Montagu thought otherwise. He was a humanist, essentially, and there was little that seemed as human to him as our skin. *Touching* was focused on what Montagu called "the mind of the skin"—how the experience of touch affects who we are and who we become. Montagu's speculations were the genesis for a lot of subsequent research on touch, in part because he speculated about seemingly everything that could be researched. He was omnivorous and *Touching* ranges so far afield that the book sometimes feels as if it is about to fall off the edge of the known world: "Extensive inquiries over many years yielded only two cultures in which mothers sometimes washed their young by licking"; "one of the earliest serious discussions of nudism and the disadvantages of clothes . . ."; "it is well known among experts and dairy farmers that hand-milked cows give more and richer terminal milk than machine-milked cows."

Then there's the mishmash. Montagu, overly confident in his internal compass, was happy to blindly theorize about anything. Large sections of the book are time-locked and painful to read today: his insistence that "the failure of mothering is the principal factor in producing the conditions which lead to SIDS," for example. But his more outlandish claims for touch are useful. They don't tell us much about how touch works but they tell us a lot about how we want to think touch works.

In this way, Montagu marked the path for nearly everyone who wrote about touch after him. In *The Vital Touch,* a recent book on the role of touch in development, the sense is said to possess almost miraculous powers. It amounts to destiny. "Can you see yourself weaving in and out of vines, up and down ledges, around and through riverbanks in search of daily food?" the author, Sharon Heller, asks:

Or shopping in the mall while effortlessly balancing an enormous bundle on your head and a baby on your side or back, as do the poised barefooted African women as they amble though the marketplace? Or fetching a coconut for breakfast, as did a Tahitian native for me, by nimbly climbing a coconut tree with bare feet and the speed and vigor of a squirrel?

If all this seems a bit daunting, you were probably not swaddled as an infant.

Mystery solved!

The purpose of this sort of story, and there are many more, is not to explain the role of touch in development. It is to exalt touch, to anoint it, to single it out as a magical elixir. Over the years, almost every aspect of child development has received this treatment. Touch is the monocausal explanation of the moment. The irony, of course, is that touch was once the monocausal explanation for why children turned out poorly: they were being touched *too much*.

It's not a surprise that we're attracted to reductionist rationales for why children turn out the way they do: parenting is a bewilderingly complicated enterprise, and if there were really one thing that made sense of it all, the whole business would be so much more humane. I feel that way; almost everyone must. What's surprising is that so many experts are willing to provide these radically simplified explanations, or at least willing to tell parents stories that clearly lead to these conclusions. The way in which Heller's book is irritating isn't unique; everyone who writes or talks about infancy is irritating in the exact same way—myself included.

Think back to the end of the section on smiling, where I cite a study on levels of anticipatory smiling—basically, more smiling at a young age successfully predicts social competence at an older age. It's a good study, rigorous and unexpected and powerful. That's why it ends the section: it suggests the deep importance of prosocial development. But it is hard not to shrink its conclusion down to a size—more anticipatory smiling *good*, less anticipatory smiling *bad*—that will get you pretty freaked out about how much your hypothetical kid is or isn't smiling.

Infants are like teenagers: you're so desperate for any insight into their lives that the littlest bits of information loom very large. By highlighting that small study, I make it—a modest contribution to a vast literature on social development—into a mountain. Suddenly, smiling matters more than anything else in infancy.

You'd expect academics, people who are by training comfortable with complexity, to be the most resistant to the idea that we're shaped by any single factor. In fact, they are often the worst offenders. Immersed in their own research, shaped by their own work, many logically see everything else as a natural extension of it. Being an expert on a subject doesn't mean you have a broad perspective on it. It usually means your perspective has shrunk.

The philosopher David Hume described this tendency well some 250 years ago:

> When a philosopher has once laid hold of a favourite principle, which perhaps accounts for many natural effects, he extends the same principle over the whole creation, and reduces to it every phenomenon, though by the most violent and absurd reasoning.

It's not a new problem, in other words. But it has a particularly pernicious effect on discussions of child rearing. Over the course of the last century, modern parenthood became an extended appeal to authority: expert knowledge replaced and superseded personal knowledge. It still does. But today there is even more expert knowledge. It's no longer limited to the grand pronouncements and weighty manuals of the child-care gurus. Now every new scientific study relevant to any aspect on child development achieves brief but immortal fame on the Internet before being filed away by Google for late-night scrolling by anxious parents. The relevant subject in each study is different—bilingualism, breast-feeding—but it *always matters a lot*: the authors will testify to the grave significance of their study. But that significance is within the discipline. It means a lot less in the messier world outside.

To almost any parent, though, it feels safer to be too credulous than too skeptical. Faced with the highly improbable task of child rearing, most of us are at our most vulnerable.

Hunting and Gathering in Suburbia

There's an atavistic streak in these appeals to child-rearing authority. We want to know how we are supposed to take care of our children—how we're *meant* to take care of them. This is an unanswerable question, of course, which is why there are so many answers to it. But if you squint at the question, it blurs into something almost answerable: How did we *first* take care of our children? Back when humans began to be humans—way back, back before blogs had comment sections for readers to offer helpful corrections—what was parenting like?

That's a question for anthropologists, not child-care gurus, and a lot of them have hypothesized that the answer might be found in the structure and daily routines of the dwindling number of hunter-gatherer cultures in central Africa. These are the lone examples of people still living in the environment in which we evolved.

Among the few surviving hunter-gatherer societies, or those whose traditions survived until very recently—the Hadza in Tanzania, the Mbuti of the Congolese rain forests—the most studied may be the Kung people of the Kalahari Desert. (They're also called the !Kung, with the exclamation point signifying a click sound, although I've left it out here.) Known more broadly as the Bushmen or the San, the Kung maintained a nomadic, foraging lifestyle well into the last century. In short, they were something out of an anthropologist's most vivid daydream, and beginning in the 1960s, seemingly every detail of their lives was tabulated by visiting researchers. Among them was the young anthropologist Melvin Konner, who spent his time among the littlest Kung. The unique importance of infancy among the Kung, he wrote, was that they may be "representative of a group of societies resembling, in their basic subsistence ecology, the original human sociocultural form." If we want to know what sort of parents we are meant to be, we should study the Kung.

Konner's fieldwork on mother-infant relationships was revelatory: it described a society in which infants were carried almost constantly and breast-fed on demand—every fifteen minutes, famously. Babies slept alongside their mothers at night; they were never left to cry. The Kung mothers carried their babies in a sling that rested on their hips and allowed for almost continual skin-to-skin contact. In the few first months after birth, a Kung infant is against his mother's skin more than 70 percent of the time he's awake; he is against someone's skin 90 percent of the time. (Infants in industrialized countries are against someone's skin between 10 and 20 percent of the time.) The Kung mothers acted in the same way our hominid ancestors must have: they kept their babies extremely, almost unbelievably close. The Kung even refused to put a baby who was awake down on the ground—they believed it harmed motor development.

Konner's work seemed to establish that there is a different model of infancy, and a different model of parenthood, in hunter-gatherer cultures. If you have a vague sense of what infancy *should* be, in the primordial sense, it was shaped by Konner's fieldwork:

> When not in the sling they are passed from hand to hand around a fire for similar interactions with one adult or child after another. They are kissed on their faces, bellies, genitals, sung to, bounced, entertained, encouraged, even addressed at length in conversational tones long before they can understand

words. Throughout the first year there is rarely any dearth of such attention and love.

Who could not be bewitched by that life? How enchanting a description of infancy! Who would long for the womb when you could have *that*? Konner wasn't writing a New Age tract: he was a Harvard anthropologist who later became a medical doctor; he remains among the most insightful thinkers on childhood that we have. His work with the Kung is legendary and it is nearly impossible to read his descriptions of their life without feeling a rebuke: their kindness and patience are so extravagant that your own parenting suddenly seems grudging and miserly.

But Kung parenting was practically negligent compared to what Jean Liedloff had in mind.

A self-fashioned anthropologist, Liedloff spent several years living in the Amazon with the Yekuana, an isolated people with age-old traditions. Her manifesto-meets-ethnography, *The Continuum Concept*, published in 1975, was an indictment of Western society in general and Western parenting in particular. It cast Yekuana society as a vision of paradise—the less-than-subtle subtitle was *In Search of Happiness Lost*—and it would become a cult classic.

In Yekuana culture, Liedloff explained, a newborn was not sheltered or separated from society but swept up in it. This total immersion in a society from birth prevents alienation and disillusionment later in life, she argued, and it begins with the mother: the feedback an infant receives from the mother is so crucial that she should be in constant contact with her baby for at least the first six months. Constant contact means constant contact: she should *never put down the child*. And if she did, then there was this: "The missing experiences of the in-arms phase, the consequent gap where his feeling of confidence ought to be, and his ineffable

state of alienation will condition and influence all that he becomes, as he grows up around the rim of the abyss where his
sense of self has been stunted."

This is the logic of *The Continuum Concept* as a whole: Liedloff
adopts a sensible stance—we should incorporate our babies into
the rest of our lives—and then makes pronouncements so large,
and so dire, and so unsupported, that you end up skeptical about
even her sensible stances. In a particularly strange passage, she
channels what life is like for a typical infant in Yekuana society
and a typical infant in the Western world. It's some twenty pages
of speculation that never admits to being speculation. Things do
not go well for the Western infant. (Sample sentence: "When he
awakens he is in hell." Bonus sample sentence: "Usually he simply
holds his thumb in his mouth against its unbearable emptiness,
the eternal loneliness, against a feeling that the center of everything is somewhere else.") Cut off from his mother's arms, the
Western infant's sexuality is distorted: he becomes, in an uninten-
tional echo of the hysteria over thumb-sucking, a compulsive
masturbator.

For Liedloff, the burden of parenthood is no burden at all: "It
would help immeasurably if we could see baby care as a nonactivity. We should learn to regard it as nothing to do." If a mother's
job prevents her from carrying her baby constantly, then she
should "give up the job in order to avert the deprivations which
would damage the baby's entire life and be a burden to her for
years as well." It is perhaps not surprising that Liedloff herself,
who told mothers that child care was so little work that it was effectively *a nonactivity*, had no children.

Despite not having children, Liedloff does not hesitate to tell
other women what to feel—what they *will* feel. If a mother carries
her child constantly, she will experience a wave of well-being
that will override all her other desires: "She will not *want* to put
her baby down." And if she does, then she's suffering from false
consciousness—she's wrong. A strident opponent of prescribed

ideologies for child rearing, Liedloff never realized she was writing a prescribed ideology for child rearing.

Liedloff's ideas were new, but her sense of authority was not: there's a deeply ironic brotherhood between Liedloff and John Watson, the arch-behaviorist. It's not in their ideas, of course: no one's philosophy is less like Watson's than Liedloff's. It's in their voices: bizarrely self-confident, despairing of everyone else, absolutely final. A thread of millenarianism runs through them: the belief that all is nearly lost—but those who follow me shall be saved.

Watson and Liedloff are extreme cases, but a hint of the end times, in their secular incarnation, lurks in almost all guides to child rearing. It has to be there: the implicit appeal of any respectable child-care authority is that he or she is saving you from purgatory. After all, if there isn't a purgatory to be saved from, what are you so concerned about? Why are you consulting a child-care authority, anyway?

That's why the past—the deep past, the original past—is a comforting place to find answers. It solves the problem of which authority to rely on: way back then, there was only one.

The Continuum Concept isn't an artifact of the 1970s. It remains in print; its devotees call it "mind-altering" and "life-changing." Any author would kill for its Amazon reviews. "Liedloff's work earned her a hallowed place as the messenger of a certain truth about the best, most instinctual way to parent," as the sociologist Chris Bobel notes. *The Continuum*'s ideas—sanded down to be more modest, less raw—have surfaced in the attachment parenting literature: many of Sears's prescriptions for parents are rooted in *The Continuum Concept*.

Meanwhile, the Kung have leapt from Konner's research to become a symbol of all that the rest of us have done wrong—a living repository of lost parenting knowledge. "What's their secret?"

Harvey Karp asks in *The Happiest Baby on the Block*. "What ancient wisdom do the Kung know that our culture has forgotten?" An Australian woman reflecting on her experience of raising children writes, "At every stage of parenting of my babies, I thought, 'What would a Kung woman do?'"

You can trace that impulse—*what would a Kung woman do?*—back through the last century: as long as we have been taking notes on how other peoples parent, we have been looking for help. The anthropologist Margaret Mead, from her earliest work in Samoa and New Guinea, saw the myriad variations in child rearing as a series of natural experiments, experiments which would solve the toughest problems of Western parenting. Mead studied the Manus children of New Guinea, for example, who grew up almost without parental interference; if Western parents wanted to figure out if "permissive" child rearing was successful, the Manus could tell us. Looking to other places for help—for *help!*—isn't a recent distortion, in other words. It's the original idea.

As psychologists paid more attention to culture, and anthropologists paid more attention to childhood, that idea filtered down into popular baby books. Ethnographic anecdotes metamorphosed, from their academic roots, into object lessons for Western parents. Many of those lessons were about touch: in the broad spectrum of parenting strategies, the Kung and the Yekuana represent the apotheosis of touch. There is simply no way to be more in contact with a baby than they are. Their stories arrived at the right time: the 1970s. After decades during which Americans had worked hard to resist the temptation to touch their children, the advice had swung back the other way: you were now *supposed* to touch your child. The equipment made this clear: strollers were out; slings were in. The old thinking: any amount of touching was too much. The new thinking: no amount is too much. Or alternatively, and more anxiously: any amount is too little.

Today, refracted through various books and ideologies, the high-touch, deep-past vision of parenting is seen as a solution to the persistent problem of what to do with this new tiny human. It's no wonder. It is hard to read about the Kung without feeling that their attitude toward infants is somehow more "right" than our own. (In fact, the "rightness" of carrying infants has been empirically confirmed: in an experiment, carrying a baby for three extra hours a day significantly reduced crying and early evening fussiness. It doesn't cure colic, though; it isn't magic.)

But the Kung style of child rearing is stubbornly embedded in the Kung culture itself. The entire structure of a Kung community supports the (many) demands of Kung parenting. A Kung mother is virtually always around other adults, who take turns holding the child. This situation is the polar opposite of that of many American mothers, who can feel marooned on an island with no one but this ferret-like creature around. As Konner observed, such a rich social environment is what makes the very indulgent Kung parenting model "emotionally *possible*" for mothers. This is the crucial point: no village, no child. It's also the point that is almost always glossed over. The Kung model of parenting transplanted to our suburban, technological society, where the television is what counts for company, is no longer the Kung model of parenting: a Kung mother herself wouldn't be able to follow it. A modern American parent hardly has a chance.

The Kung solution to crying isn't any easier to adopt. When a Kung infant begins to whimper, an adult usually responds within fifteen seconds—almost instantly. But an adult does not necessarily mean a parent. Almost half the time a Kung infant cries out, he is comforted by someone who isn't his mother or by his mother plus someone else. When the mother responds alone, other people offer to take the child later on. The Kung mother isn't abandoned with a wailing infant. But despite this shared caretaking, the Kung, as Konner notes, "have often been misrepresented as having almost exclusive maternal care."

It's worth dwelling on the distinction between exclusive maternal care and alloparenting—the term for when someone who isn't a parent acts as a parent, as the Kung do when they respond to *any* crying baby. If the most important messages to get across to a baby—love, security, commitment—are communicated through touch, then the obvious follow-up question is: does it matter who's doing the touching? The parent or the alloparent?

From the perspective of attachment theory, all child rearing is aimed at the same end: the tight bond between mother and child. There aren't multiple different strategies toward a successful outcome—there's only that one. (Bowlby waffled on this a little bit but not much: his hypothetical caregiver was clearly a mother.) The mother is supposed to be doing the touching. This argument wades into the evolutionary past for evidence—the low fat content of human milk, for example, which required infants to nurse frequently, for which they needed a mother right there, all the time. In devising his theory, Bowlby cited the behavior of primates like gorillas and chimpanzees, for whom child care is exclusively maternal—no one else need apply.

But studies of hunter-gatherers like the Kung, the very people you'd expect to be closest to our deep past, have shown that caregiving by someone in addition to the mother is common, even if other people rarely supplant the mother as the primary attachment figure. The amount of alloparenting varies widely, but the existence of it is the rule, not the exception. And as scientists learned more about primates, Bowlby's conclusions were undermined: fully half of all living primates do not provide exclusive maternal care.

As more research of hunter-gatherer cultures was published, a pattern emerged. For the Efe, who live in the northern Congolese rain forest, alloparenting is completely ordinary. Up until toddlerhood, an Efe infant rotates among multiple caregivers several times during a single hour; she nurses from multiple women. Even when the mother is present, she isn't necessarily the primary caregiver. Alloparenting is a cushion against the excruciatingly high mortality rates of the Efe: the more alloparents an infant has at a year old, the more likely she is to still be alive at age three.

Among traditional societies that are not hunter-gatherers, alloparenting is no less unusual. In West Africa, Beng mothers return to physical labor in the fields when their infant is only a couple of months old. How do they manage this? They hire someone in the village, often a young girl, to carry the child for part of each day. But because such a girl is usually only available part time, any Beng mother has a long list of sitters who can fill in. "Given frequent changes of caretaker," writes the anthropologist Alma Gottlieb, who lived with the Beng, "it was not rare for a mother to be unaware of where her baby was, and in whose care, at some points in a typical day." According to Gottlieb:

> A mother may hand her baby to her first-morning baby holder knowing that the latter is likely to pass the baby to another person if she herself becomes tired or if the baby fusses or if another person requests the child. By the time the infant is brought back to the mother to breast-feed—depending on the child's age, this might be up to a few hours later—the little one may have been passed around to several people as caretakers. The mother may not even hear the full list of who was taking care of her child during this period.

What do we make of all this extra-maternal care? The psychologist Edward Tronick, who has studied the Efe, argues that the whole idea of a "living" evolutionary past is a fiction. There isn't a more "natural" way of life, Tronick says. "Biology is no more the destiny of the Efe than it is for us." Instead, he says, the Efe philosophy of child care is just an adaptation for their environment: "These adaptations are neither more nor less biologically based than those of other cultures. That is, the Efe lifestyle is no more or less genetically based than the lifestyles of other peoples."

For Tronick, there isn't *an* answer to the question, How are we meant to take care of our children? There are many answers. "Our decisions about child care practices are really decisions about cultural values: about what we want our children to become."

The Kung aren't a time capsule of *Homo sapiens* parenting. They're a time capsule of parenting *in the Kalahari Desert*. If you are in search of parental wisdom, this is bound to be disappointing. It is extremely unlikely, after all, that you too live in the Kalahari Desert. A few academics have written that the longing for the "original" mode of parenting is a parochial, patronizing idea—it insults the complexity of the age-old cultures that it claims to venerate. That's true, of course. But there's a less academic, more boring objection, too: we don't live in the Kalahari Desert. Or the Amazon. Or the Congolese rain forest.

Margaret Mead's hope—that the many cultural variations in child rearing would be a tool kit for Western parents to use—suffers what might be called the Kalahari Desert problem: the fact that all those variations evolved in their own cultural context. Outside of that context, they're meaningless or dysfunctional or worse; at a minimum, they're frustrating. It's puzzling that Mead of all people convinced herself otherwise: when it came to child rearing, she was a cultural anthropologist who somehow forgot about culture.

In this omission, she was way ahead of her time: many decades later, culture is what always gets erased from the practicalities of parenting. No parent tries to emulate hunter-gatherer societies in any other sphere of life: for sustenance, we do not go foraging instead of grocery shopping. But with our children, we start from the premise that all things are possible and that parenthood is the only relevant fact in the world, the shared experience that overrides all differences. Our child allows us access to the Amazonian within. But the choose-your-own-culture version of parenting has a stubborn problem: no parent is a culture.

The Search for a Grand Unified Theory of Babydom

We have looked to the deep past to solve the stubborn problem of how to raise our children. We have looked almost everywhere else, too. We have looked to sheep, even. And sometimes we find answers there: there were many moments over the last century when someone decided that he'd stumbled across a solution—*the* solution. Harry Harlow thought he had: he thought that touch was the solution. He'd expected the monkeys raised by cloth "mothers" to turn out beautifully. They didn't: they turned out debilitated. Wowed by his insight into the importance of touch, Harlow only later realized what the cloth "mother" didn't do. "She didn't teach, direct, or steer the baby toward others," Harlow's biographer Deborah Blum writes. "From cloth mom, baby really learned more isolation, separation from others."

Harlow had revolutionized the idea of what infants needed: he'd proved something that seemed too warm and fuzzy to be provable. But Harlow's baby monkey, despite being "touched," was still the bubble-baby of John Watson's era. "The original concept of his surrogate mother," Blum writes, "was still grounded in that sterilized notion of a healthy baby—clean, fed, warm, disease free, isolated from harm. Once again that had been proved incomplete at best and destructive at worst."

Incomplete and destructive: these are the poles that singular child-rearing solutions—touch! don't touch!—oscillate between. No one gets a better grade than an incomplete. These solutions never have much to do with the child. They have to do with the adults. The psychologist Jerome Kagan once wrote that although many people a century before John Bowlby would have agreed with his ideas, "few would have written three books on the theme of attachment because, like the blue of the sky, the idea was obviously true." (A *half* century before, though, it wasn't.) Ideas only catch fire in incendiary conditions. Bowlby's ignited, Kagan contends, because it upheld something a dizzy, disconnected society needed to believe in—the sacredness of the mother-child bond.

The sucking hysteria—a singular solution of its own—was the creation of a stern new science of child care devised for a strange new world. Harlow's ideas were a revolt against a behaviorism that divorced affection from parenting. The great sucking menace now seems inexplicable and alien, but most of us regard touch the same way Harlow and Bowlby did—as utterly essential. There are very good, very basic reasons for thinking of touch this way. As the neuroscientist Robert Sapolsky, who studies stress, puts it:

> Touch is one of the central experiences of an infant. We readily think of stressors as consisting of various unpleasant things that can be done to an organism. Sometimes a stressor can be the failure to provide something essential, and the absence of touch is seemingly one of the most marked developmental stressors that we can suffer.

This is as close to indisputable as anything in this book. And it means that, at least in the near future, we're not going back to the days when hospitals forbade parental visits and neonatal wards enforced hands-off policies. There are many disorienting, distressing things in the history of infancy—I have left out a lot; I have been kind. But the phobia of touching—the existence of a time when you were not supposed to touch *your own child*—was, just viscerally, the hardest to bear. Each time I saw Isaiah after reading about it, I'd hug him tightly.

If the absence of touch is destructive, that means that touching is essential. But it doesn't mean that more touching is the key to the species. There are other reasons why touch, like the sucking hysteria of old, has attained Very Important Status—the sort of reasons Kagan might supply. If someone writes three books about the sky being blue, something besides the sky is the real subject.

In any case, we shouldn't get cocky about our certainties. The history of how we think and talk about babies suggests that all of us—the parents, the experts, me, you—are sunk too deep in the present day to see beyond it.

Harry Harlow was still living in John Watson's world. We're still living in John Bowlby's. A century from now, though, whoever's around won't be. They'll have a whole new set of anxieties they'll need to be reassured about.

14

Finding the New World

Until now, the adults have hogged all the screen time. The baby has been a bit player in all this: he doesn't touch; he is touched. But we have reached the pivotal point in the plot when the minor character takes over the movie.

A new baby doesn't touch with the body part that you'd use for touching. She has a mind of her own: she substitutes another part altogether. Babies use their mouths for touching—specifically, they use the inside of their mouths. If you give a newborn a toy, she will invariably bring it to her lips, and when the psychologist Philippe Rochat gave newborns pacifiers with different shapes, he discovered that the more eccentrically shaped the pacifier, the more the infants used "nonrhythmical, disorganized movements of the tongue, lips and jaw." They didn't suck on the pacifier; they tried to *feel* it. And they could: the infants were able to visually identify a pacifier they'd never seen but had sucked on. The perception crossed their senses: the infants had effectively "seen" the pacifier when it was in their mouths.

But someone has to give the baby the pacifier—no newborn would be able to reach it on her own. Babies are cataclysmically bad at reaching. If you were to track the minutes, which no sensible parent would be foolhardy enough to do, the first few months of parenthood would mostly amount to: 1) watching your child try, and fail, to fall asleep, and 2) watching your child try, and fail, to reach things that look really fun.

When Isaiah was four months old, a friend brought him a nifty new toy—the sort of toy with so many colors and textures it seems designed to induce seizures. She held the toy out for Isaiah and his eyes went wide. He batted at it with all his fifth-percentile-weight might. He missed. He tried again. He missed again.

This is the sort of thing parents are not supposed to say, but nevertheless: there is something inescapably pathetic about someone, even an infant, who is unable to make contact with an object in front of his face.

"You know," the friend said, "I think he'll figure it out soon enough."

She was right. He did. *This was the last thing I expected.* T. Berry Brazelton used to reassure the parents of obsessive thumb-suckers by saying that he hadn't seen many adults who still sucked their thumbs. I'd never seen a healthy adult who was unable to reach—but maybe I just hadn't noticed. Maybe there was a whole group of people not salting their food because they couldn't reach the shaker.

We think of infancy as a time of tremendous achievement, but what dominates the daily life of an infant is failure. To be a baby is to live out Beckett's dictum: "Try again. Fail again. Fail better." Babies never stop trying: they fail better. At any point in life after infancy, such a success-to-failure ratio would be intolerable.

But even though they fail to get the toy, babies seem to find satisfaction in the attempt itself. It's akin to their fundamental orientation toward the social world—the way babies delight in conducting a conversation of nothing but *coos.* Just as babies want to socialize for socialization's sake, they want to move for movement's sake.

With any luck, this is some compensation for the sheer frustration. A baby who's learning to walk has at least the brief experience of walking *before* falling. The first successful reach is preceded by months of misses. The misses begin almost at birth: when infants too young to support themselves have their head and torso stabilized, they will make motions called *prereaching*—they stretch their hands and arms toward a toy. Even newborns, who are pretty much a floppy mess, do this. They always miss the toy, but their arms slow down when they're near it—the toy is what they're after.

It is only after many tiny, unappreciated advances that the infant gets the toy. The act of reaching turns out to be astonishingly complex. According to the psychologist Claes von Hofsten, reaching depends on only—only—this: "differentiated control of the arm and hand, the emergence of improved postural control, precise perception of depth through binocular disparity, perception of motion, control of smooth eye tracking, the development of muscles strong enough to control reaching movements, and a motivation to reach."

It is also the very first catch-22 in life: the very act of reaching is what throws off the act of reaching. (Take a deep breath.) When you reach for an object, you move your upper body—which means that the object is now in a different place, relative to your body, than it was before. Your original calculations, made before you began reaching, are already incorrect. Those calculations have to be recalculated, and then recalculated, until finally—you arrive at the thing itself. But not only are the calculations flawed, they were calculated with flawed data: when you reach, your head wobbles, which means your eyesight wobbles, too.

It's a dizzying, compounding array of difficulties. Because the baby has moved, the object she's reaching for has effectively moved, too, but she has no idea where it is, because in moving she's lost her ability to track it.

But the hardest part of reaching isn't actually reaching. It's posture. Babies will spend long stretches of time reaching for multicolored jungle animals suspended above their body because the position solves the problem of posture—the floor steadies their body. Being able to steady their body on their own—to sit up—comes far later, around the half-year mark. The development of independent sitting has been called, appealingly, "one of the premier accomplishments during the first year of life." But that may even sell it short. The demands of sitting are cruelly paradoxical: in order to sit up, you have to have already been sitting up—the muscles supporting the sitting position do not develop until a baby *starts* to sit up. A baby who's trying to sit up lacks the very foundation for what he's trying to do. It's like learning calculus without knowing algebra.

This level of detail seems absurd because we know how to reach: the act feels so natural and smooth to us. And because reaching feels so intuitive, it is impossible to contemplate what being *unable* to reach feels like.

At some point, infants realize that actions aren't a random series of gestures. When people act, they have some goal in mind: they *intend* to do something. This insight makes sense out of the world. Actions are motivated by intentions: without that knowledge, the workings of everyday life would remain a wholly inscrutable mystery.

This is the sort of idea that I find staggering. Not the idea itself. But that infants *have to have* the idea. If there is a moment when they realize that actions are more than Jackson Pollock–like gestural swirls, that means that before they figure that out, *they don't know that.*

In recent years, some historians, rather than explaining what happened in the past, have tried to capture what the past felt like. They have tried to unearth the sensory past: the smells, the sounds, the sights. That's really hard to do, for the historian and the reader: once you've heard recorded music, you can't imagine not having heard it. Writing about infancy presents a similar problem. You can get at the science behind infancy; you can get at the history of it, the cross-cultural studies, the medical case reports. But you can't get at what infancy feels like: once you know people have goals behind their actions, you can't imagine not knowing it.

We were all babies, but it may be easier to imagine Elizabethan England than infancy. Sometimes, though, we can see when babies make the leap to understanding things in the same way we do—when they become like us.

The experience of reaching triggers a leap of this sort. A few years ago, some psychologists had the idea of having infants wear "sticky mittens," mittens covered with Velcro that allow the infants to pick up toys covered with Velcro. The babies don't have to have mastered reaching; they can just swipe at the toy. The psychologists studied babies who were three months old, an age at which infants are lousy at reaching and at understanding that actions are goal-oriented. But if the infants had the experience of reaching, would they understand what other people were doing when they reached?

The researchers used a habituation paradigm, an experimental design common in infancy research: if you show infants an event enough times, the infants get bored—they become "habituated" to the event—and look away. If you then show a different event and they suddenly pay more attention, you know they have perceived the new event as novel. They register the difference.

In this experiment, infants watched as a person wearing a sticky mitten reached for different objects. The infants who watched the person before they wore sticky mittens themselves failed to register anything different when the person reached for a new toy. But the babies who *had* played with the sticky mittens clearly registered the difference. In fact, the more experience babies had with the sticky mittens, the more strongly they registered the difference—they paid even more attention to the fact that the person was reaching for a new toy.

With amazing speed, the act of successful reaching—of attaining a goal—changed how the babies understood the world. *Doing,* not just seeing, was what mattered. What an infant gets when he reaches for a toy, it seems, is far more than just the toy. He gets something more metaphysical: a lesson in what it means to have intentions. And the realization that everyone has intentions of their own—that everyone around him is constantly reaching for their own toys.

If being touched is essential to human development, and there's every reason to believe that it is, then the act of touching gives infants a crucial sense of what it means to be human—to act as a human. From passive to active: it is an enormous, mind-straddling change of perspective.

For the infant, though, the new toy itself may be a lot more exciting.

Part Four

TODDLE

Learning to walk was the dividing line between the uncertainties and vulnerabilities of the first year of life, when even the very nature of the child was suspect and disquieting, and the youngster's entry into the family as a functioning junior member. After that achievement, the child was simply expected to blend into the everyday life of the family as best it could. The age of special needs was over.

—Karin Calvert, *Children in the House:*
The Material Culture of Early Childhood, 1600–1900

I can't remember exactly why I thought it might be interesting to put a baby on a treadmill.

—Esther Thelen, "The Improvising Infant"

15

Who Put the Norm in Normal?

Labor Day weekend, Isaiah's first year, Anya's parents in epicenter-of-nowhere rural Massachusetts. I remember that the weather was still summer, and the garden was in full harvest, which in New England means mainly kale. Isaiah's head was the size of a late-summer cabbage: we have a photo from above; the spheres are pretty much one-to-one. I also remember that on Monday, Labor Day itself, when we were sprawled on the back porch, emphatically not laboring, Isaiah started to crawl.

He hadn't been stationary for months at that point. He'd squirmed and rolled and inched and scooted. He'd required baby-proofing. But he hadn't crawled until, all of a sudden, that afternoon, he decided it was time. It felt like a conscious decision. Everyone's here. Why not just do the thing now? Hands, check; knees, check: *go*.

Isaiah was born in early December. This was early September. He was, quite frankly, a little late in deciding to get around to it.

How did I know he was a little bit late? Arnold Gesell told me. He told you, too.

Born in 1880 in a tiny Wisconsin town on the Mississippi—a

two-street town, he called it—Gesell rose to become the most influential child-rearing expert of the first half of the twentieth century. A pediatrician and psychologist, he published the first timetables for "normal" child development. He established the concept of developmental stages. He created a pioneering system for diagnosing developmental problems. Initially reluctant to write for the mass market—his version of a parenting manual, *Infant and Child in the Culture of Today,* only appeared in 1943, after he'd logged decades in academia—he changed the genre when he did: *Infant and Child* sparkled with epigrammatic, lively sketches of the growing child. In her history of modern American child rearing, Ann Hulbert credits it with "pioneering a new form of anecdotal lore that has typified the child-centered branch of advice literature ever since."

Today Gesell is largely forgotten, remembered mostly in citations to his work on developmental norms. But he laid the foundation for how we think about child development: beneath all our assumptions about how infants should behave, and especially about how they should move, is the work that Gesell began a century ago, a few blocks away from where Isaiah would be born.

Long before words arrive, locomotion is the preeminent achievement of infancy. It arrives in fits and starts and falls, until, suddenly, your inert newborn is pulling out the kitchen drawers and clambering up into the sink for the butcher knife. It is stubbornly hard to believe that the toddler who hurtles his way across the living room used to be a blob who could only ooze his way across his mother's chest. How did this happen?

For many decades, that question was answered like this: all babies follow a steady predictable path from birth to walking, a path that's neurally dictated—the brain says do this, so the baby does it. The path is marked by a series of universal developmental stages and norms. No healthy baby wanders off it.

But in recent years, psychologists have offered a very different answer: babies discover movement on their own—learning to move, far from being generic, is a creative act. And a cultural act: babies in Kansas learn to move differently than babies do in Kenya. And it changes over time: babies today begin to move differently than babies did a few centuries ago. Rather than being unvarying, learning to walk is confoundingly variable. We all end up walking. But none of us get there in exactly the same way. We never have.

But the earlier answer—the promise of *the one true neural path*—is still sacrosanct. Developmental norms too vague to be helpful still have the power to terrify parents. Pediatric offices are still decorated with posters and charts that show erroneously predictable accounts of motor development.

The story of toddling—how babies go from oozing to hurtling—offers us gratuitously thorough evidence of just how provincial our knowledge of infancy is. Not knowing much isn't a condition unique to us. It's the fate of all parents. It's also our salvation: having a script to follow, any script at all, is what makes parenthood possible. Otherwise we'd be paralyzed ten times a day.

But we tend not to see our provincialism for what it is. We have disguised it as objective truth, even though today's objective truth is often, somewhat disconcertingly, different from yesterday's objective truth.

This transformation of cultural belief into scientific fact begins, as much as it can begin with anyone, with Arnold Gesell.

The Unbearable Normalness of Infancy

Before there were norms, there were numbers. The impulse to measure a baby is now so ingrained that it seems to exist outside of culture—it feels more like instinct. But the basic metrics of childhood—how much does he weigh, how tall is he—were taken down for the very first time during the Enlightenment, amid a blizzard of data.

Measurement was a novel enough idea that, after a friend of the French naturalist Georges Buffon tracked his son's growth, Buffon found the numbers notable enough to publish, without further elucidation, as if they were the record of a rare species. Not until 1787, when the German philosopher Dietrich Tiedemann published a detailed study of his son's first thirty months, was infancy measured with any real rigor. No one had bothered to chart the explosion of growth in infancy for a very simple reason: the idea hadn't occurred to anyone. Before children were to be observed, they had to become worthy of observation.

Tiedemann was ahead of his time. To most educated men, it was the rise of evolutionary thought that made babies worth the trouble: after *On the Origin of Species,* the infant was seen as a portal, a way to peer back through evolutionary time. That's why he was worth observing: if you looked carefully enough, you could see nothing less than modern man emerging. Darwin, of course, kept a diary of his son, twenty years before *Origin* was published. Suddenly, the potential observations were endless. Societies for child observation were formed across England and the United States.

All this observation and measurement became slowly systematized, taken out of the notebooks of amateurs and inscribed in the charts and monographs of academics. In 1882, the scientist William Preyer published *The Mind of the Child,* a tour de force collection of strict, thrice-daily observations of his son and other children from birth to age three. In part, Preyer established the tradition Jean Piaget followed in observing his three children: he simply *watched* them. To test his theories, Piaget also ran ingenious miniature experiments, but what he did first was observe. His grand theories were built on the observations he'd made with his own eyes.

Arnold Gesell's work implied the opposite: that observations made by humans, no matter how smart or sensitive, couldn't be trusted. Humans were ruinously subjective. Gesell's work was the logical endpoint of the rise of raw metrics. Only the camera told the truth.

Arnold Gesell was named director of the Yale Clinic of Child Development in 1911. Almost immediately, he installed a photography lab and a film department. A soon-to-be-famous "photographic dome" was constructed, with a crib in the middle—part testing surface, part playpen, part fort—and cameras on the interior and a one-way screen surrounding it. These were telling priorities. "Cinema analysis is a form of biopsy which requires no removal of body tissue from the living subject," Gesell would write. "It permits us to bring this behavior without any deterioration into the laboratory for searching dissection. The dissection is equivalent to a microscopic examination of the histology and the function of an organ *in vitro*."

With the trove of images he accumulated, Gesell published the massive *Atlas of Infant Behavior,* a minutely detailed canvas of babydom. In two bullet-stopping volumes, it sketched a map of seemingly every movement a baby could make, or as Gesell put it in insistent italics, *"a systematic collection of specimens of infant behavior patterns."* With some 3,200 photos of babies hard at play, it was a book custom-built for the new scientific era: any parent could skim its photo spreads and marvel at the wonder of carefully controlled development. The orderly progression of growth the book charted was so clear, and so satisfying to observe, that it seemed beyond objection.

Frozen on film, the exact components of behavior could be dissected in a level of detail inconceivable before. From the *Atlas,* "Behavior Situation: Standing, 8 phases; Age 60 weeks; Child G44": "4.75. Flexes right arm, upper arm slightly abducted, forearm directed forward and downward at lower chest level; opens both hands; flexes right leg, lifting foot 1 inch and moving it forward . . ." The narrative continues for nearly a hundred more words, before marking the time again: 5.75. A mere single second has elapsed. From "Cup and Cubes; Age 40 weeks; Child: G47": "Bangs left-hand-cube against bottom [of cup]. Releases handle from right hand. Bangs left-hand-cube against bottom . . ." You have the feeling this particular behavior situation could go on for some time.

Gesell's theory was inseparable from his technology: each fraction of a second, each element of a movement, could be isolated and preserved and identified. In painstaking detail, Gesell laid out twenty-three different stages of prone behavior that an infant must pass through, in the correct order, prior to walking. There were fifty-three separate stages of rattle behavior. Nothing was more Gesellian than the impulse to arrange behavior into stages. By the end, he established a series of sequential developments that began at four weeks old and only petered out, finally, at sixteen years.

Age was Gesell's defining metric. "We think of behavior in terms of age, and we think of age in terms of behavior," he wrote. If this makes intuitive sense, it is because what Gesell was proposing, we now assume. "It is probably hard to overestimate," as the psychologists Esther Thelen and Karen Adolph have written, "how thoroughly we have internalized the idea of age-appropriate activities as an index of intrinsic biological functioning." In Gesell's era, this idea was still new enough to be novel. A new consciousness of time and age had emerged between the end of the nineteenth century and the beginning of the twentieth, and it was reinforced by nearly every social development: the radio and the telegraph; the railroads and the division of the country into time zones; the new gospel of efficiency and productivity. In the span of a half century, starting shortly after the Civil War, American society went from being oblivious of the clock to rotating around it: the language of being "behind" and "ahead of time" took up permanent residence in the national vocabulary. Age began to play a central role in how people thought about themselves. Birthdays became newly significant events. For the first time, there were different doctors for people of different ages, and pediatricians sorted the youngest demographic even more finely. There was not just infancy and childhood; there was now also adolescence.

Attaining the right milestones at the right age was half the story. They also had to be in the right order. Order was the supreme value for Gesell, the foundation of his understanding of infancy. The formation of bones in the wrist, he said, varied more than normal infant behavior. Gesell was deeply influenced by G. E. Coghill, an embryologist who made a landmark study of the salamander embryo. Coghill had linked a salamander's movements with its neural development: its actions followed its neurons. For Gesell, the human infant was the salamander, writ slightly larger and drier: Coghill filmed salamanders beginning to swim; Gesell filmed infants beginning to walk. When you observe an infant's behavior, Gesell said, what you are actually seeing is the infant's nervous system.

Gesell applied Coghill's embryology to mental growth, too. He proposed the first set of mental norms, sometimes called "maturity traits" or "gradients of growth." For Gesell, mental "normalness" was no more complicated to measure than physical. There were "wholesome habits of eating, of sleeping, of relaxation, and of elimination," all of which reflected the "proper organization of the nervous system. The child who is not well trained in these everyday habits has not learned even the first letters of the alphabet of nervous or mental health." There were also "wholesome habits of feeling" and "healthy attitudes of action." And all of these could be measured.

Moving from physical growth to mental growth was a daring leap, but Gesell hardly hesitated: he reduced each age to a few apposite adjectives. From *Infant and Child in the Culture of Today*: "THREE has a conforming mind. FOUR has a lively mind. THREE is assentive; FOUR, assertive."

In all his work, Gesell rarely made the most obvious observation of all, so obvious it hardly counts as an observation: from a very early age, babies have personalities. They are identifiably themselves. Children were individuals, Gesell acknowledged, but they always seemed to be resolutely *typical* individuals.

In his laboratory, Gesell observed more infants more systematically than anyone ever had. What he took away from it all was their essential sameness: some quibbling variation aside, in Gesell's work all children pass through the same stages at more or less the same times. The inexorable nature of development should be reassuring, he said. The terrible twos weren't the fault of the child or the parent. They were just another stage. In his books, Gesell redefined bad behavior as age-appropriate behavior. This too would pass, he advised parents: in the meantime, the child shouldn't be punished; the child should be tolerated. For Gesell, adopting a new parenting strategy to deal with the behavior was beside the point. It would have been like adopting a new parenting strategy for a raccoon: it was still going to knock over garbage cans, regardless of what its mother said.

This view was a frontal attack on the learning theory of John Watson, who argued that the child would do whatever his mother said—that behavior, above all, was what mattered. Gesell argued that children were bounded not by behavior but biology: you could only change them so much. You had to trust that the child knew, even if in some essentially unknowing way, what she was doing. Unlike Watson, Gesell was trying to listen to the child, not make the child listen to him. It was a paternalistic sort of listening, though, since he already knew what the child was going to say—the hours of footage from his laboratory had already told him.

His approach was also the polar opposite of what psychoanalysis, then seeping into American culture, advocated. In Gesell's theory, responsibility started with the child (or better yet, with the child's very biology). In psychoanalysis, responsibility started with the mother. Many psychoanalysts worried that Gesell, with his talk of inexorable and standard development, permitted parents, and especially mothers, to let themselves off the hook too easily. They need not have worried. In Gesell's writings, a background hum of anxiety about anything *non*standard was always audible. "In replacing behavioral demands with behavioral norms," as Ann Hulbert has observed, "Dr. Gesell and his colleagues promised parents relief, but also supplied them with new cause for alarm."

Indeed, Gesell's descriptions were so detailed they effectively encouraged parents to worry. In *Infant and Child in the Culture of Today,* Gesell provides a profile and schedule of the typical child at various ages: The forty-week-old "lunches at about 1:30 on spoon-fed vegetables." He "may cry out momentarily during the course of the night, without waking and without requiring attention." The eighteen-month-old plays in a "protected corner of the yard" with "only marginal supervision." After his nap, "he wakes happy and cheerful." Two and a half is the "paradoxical age" but by age three "he has himself well in hand." In *Infant and Child,* these dioramas of childhood—natural history–like displays of the child in his native habitat—are constructed for every major milestone up to the age of five.

But what if your eighteen-month-old didn't wake up happy and cheerful from his nap? Surely he was *supposed* to. Surely most children did. After all, what was the point of such descriptions if they weren't supposed to correspond to your child?

Gesell anticipated these anxieties. The norms, he wrote in italics, are not *"set up as standards and are designed only for orientation and interpretive purposes."* He cautioned that they "are readily misused if too much absolutist status is ascribed to them," a fact he knew, as Hulbert notes, "from having arrived at them by observing countless deviations." Norms should only be used by physicians and only then with great "judiciousness," Gesell warned. "The lay person should not attempt to make a diagnosis on the basis of such norms."

In an irony he could hardly have overlooked, his plea that "the lay person" should not pay too close attention to norms appeared in a book, in part about norms, that was written for the lay person. His plea was in vain and he must have known it. His norms had given untrained parents new insight into what proper development was supposed to look like. Suddenly, every mother could be an expert: she just had to look at the numbers. Every mother would now judge her newborn son, "especially in the period of infancy, by holding him up to Gesell," the pediatrician Joseph Brennemann predicted in a 1933 speech. "The contributions of Gesell are the most fascinating and instructive" in decades, he admitted, but they "are readily accessible to the laity and therein lies a certain danger. No matter how much stress may be laid on the fact that there are fairly wide normal chronologic variations in development the mother is going to accept the average as the established standard with resulting elation or depression as her child rates ahead of, or behind, that standard."

The job of the pediatrician was no longer simply to cure sickness. It was to convey what it meant to be in good health, to provide parents with perspective on how healthy—how normal—their child was or wasn't. By 1934, norms were pervasive enough that they inspired a vicious critique, *Your Child Is Normal*, by the psychologist Grace Adams. "The American craze for standardization has given us, among thousands of other standardized objects, the standardized child. It is a child who at the moment of its birth weighs exactly 7½ pounds and measures, from the tip of its straight black hair to the sole of its curled pink foot, precisely 20½ inches." It is a child, Adams went on, "who attempts to take its first toddling steps and to babble its first meaningful words on the day it commences its second year of life." Her response to such uniformity was to redefine normality as *abnormality*: "It is the very unexpectedness of their variations from the artificial norms which affords the greatest proof of their true normality."

For Gesell, though, the solution was *more* standards: if the norms are wrong, find better norms.

Over time, though, the "normal" child came to mean not the "average" child but the "ideal" child. "There was a transformation from the typical into the desirable," as the psychologists Thelen and Adolph put it, decades later. "Gesell elevated the typical child, who of course was no child, into a biological reality with profound consequences for both theory and practice."

Gesell's norms became the foundation for the Bayley and Denver developmental scales, which guide pediatricians today. In other words, no parent today can avoid benchmarking their child against a set of developmental norms that Arnold Gesell invented nearly a century ago. No parent a generation ago, or even a generation before that, could have avoided it, either. Ours, unbeknownst to us, is the Gesellian child.

Behind the Times

Isaiah sat up late. He crawled late, too, and he only agreed to walk when he was fairly sure he wouldn't fall, which was—inevitably—late. With the exception of his ability to taxonomize heavy machinery—crudely, but still—he talked exceptionally late.

He did smile right on cue, though. There were consolations.

With cheering irony, Isaiah missed almost all of his developmental milestones but met nearly every inclusion criterion of the original Gesell sample: he was a white middle-class child partly of German and British ancestry, living in a two-parent household, in New Haven, Connecticut. He was the prototypical child in almost every respect except his abnormalities.

Late, early, on time: in our what-to-expect era, when baby Web sites offer weekly, birth-date-timed developmental newsletters, you can hardly avoid knowing where your child falls on the developmental spectrum. As Isaiah fumbled through his first years, he was accompanied, via child-rearing guides and Web sites, by the phantom presence of the typical child, whose textbook development made him look like Leo the Late Bloomer.

These benchmarks followed him off the page. Until your child is past walking, milestones—ahead, behind, *nailing it*—are the unavoidable, almost unconscious conversation topic at any playground. While the children ignore each other in the sandbox, the parents practice socialization, and the obvious thing to talk about is what's at your feet: what your children are and aren't doing. There's a template for these conversations: first you ask about age, which is important because you can then ask about motor development— *Is he* [insert form of locomotion] *yet?* If you don't know the age, you can't assess the development. All this always precedes introductions: ascertaining the status of crawling or walking comes well before you learn Milo or Zoe's name.

Isaiah was almost translucent for his first year: his head and height were respectable, but his weight hovered around the fifth percentile. He had what I began to call, with defensive faux-humor, a humanitarian-relief physique: his bones were traceable under his skin while his belly ballooned outward. *For just pennies a day, you could keep this child's pants from falling off.* He was his father's son: I still have to be careful my ribs don't catch on something. He made other babies his age look like they were doping. He was, in short, the sort of infant about whom well-meaning strangers want to offer unsolicited advice. So when talking to strangers who looked like they might be too helpful, I occasionally took a month off his age, as if he were a thirty-something actress. I remember him being stuck at twelve months for a while.

In other words, in order to derail conversations about whether

I should be worried about my son, I, somewhat worryingly, lied about him. Telling lies about your child's age to total strangers may reflect a deep personal failing. But it is not mine alone. "I Lie About My Child's Age," a mother confessed a few years ago in an essay on the baby Web site Babble, an infallible guide to the post-hipster parenting zeitgeist. The author's son—seventeen months and *enormous*—wasn't walking yet, which everyone informed her was a Serious Problem. Except it wasn't: the kid had been checked out. He was fine. He was just late. So his mother rounded down his age, sometimes a little, sometimes a lot: "'Oh,' the parents would sigh, relief flooding their faces. 'That makes more sense.' Gone were the furrowed brows and awkward talk of early intervention."

It's a gentle, sweet piece. And so it was attacked in the comments as an example of brutally bad parenting: "You are attempting to cover up your own insecurity by stripping away months of your child's life. He has lived 17 months! He should be getting credit for all he has done in that time." "You're pathetic. Your inability to be 'uncomfortable' because your son isn't walking yet is all about you." "What if your son does turn out to have a real problem? Will you just continue to lie?" Then there are the ostensibly helpful comments with wholly incorrect information: "Babies who walk too early miss out on brain development," someone insists. "Skipping the crawling phase results in problems later on."

A year's worth of playground conversations condensed—some supportive, some infuriating—the comments section recapitulates why the author ended up lying in the first place: to avoid wading into the comments section.

Milestones now form the bedrock level at which we understand child development—how we tell right from wrong. In the comments section or on the playground, we all have a sense of what correct and incorrect motor development looks like.

The only problem is: that sense of whether things are correct or incorrect isn't always correct.

Crawling: Really Important!
Also: Totally Unimportant!

Take crawling—a momentous lurch forward in development. Before crawling, babies roll over; they yank their heads up; they sit humped like bear cubs. They are a bundle of extremely inefficient activity. Of all the developmental stages, the transition to crawling represents the most fundamental change, for the parents and the baby. You can't get more basic than going from *not moving* to *moving*.

In his taxonomy of prone behaviors, Gesell identified "passive kneeling" as the first prone stage (at a week of age) and itemized seventeen intermediate stages before he got to hands-and-knees crawling (at ten months). For Gesell, none of these nineteen stages were optional: the to-do lists of all babies were the same. Gesell was clear: healthy babies crawl. Today, except on the Denver Developmental Test, crawling remains a major milestone—most scales place it at eight months. (The shift from Gesell's ten months to the Denver Test's eight isn't unusual. Over the last century almost all motor milestones have arrived at earlier ages, possibly because of quality-of-life improvements like better nutrition.)

But it is possible that *not* crawling is more common than crawling. Across traditional cultures with no apparent connection to each other, crawling appears only as an aberration. In rural Turkey, infants never crawl—it is considered too dirty and dangerous. In Jamaica, over a quarter of infants never crawl, and those who do crawl so late they start walking at the same time. On the island of Wogeo off New Guinea, any baby who shows an inclination to crawl is immediately swept up by an adult—no one crawls on Wogeo. The ethnographic examples are almost endless.

The anthropologist David Tracer, who has studied the Au people of Papua New Guinea for over twenty years, noticed early on that Au babies never seemed to crawl. They were rarely even on the ground, for that matter, and on the few occasions when they were, the babies were positioned sitting up: long before they could sit on their own, they were molded into a sitting position. Au babies had almost no experience of being prone, and when they started to move, they did not crawl like Western babies who'd spent their brief lives lying on the ground. Instead the Au *scooted*, edging forward on their bottoms. If you ask the Au, scooting is the universal stage that precedes walking.

Tracer has argued that crawling is a comparatively recent development, a mode of locomotion that arose along with clean surfaces to crawl on. Before that, there were simply too many belly-level dangers, including human waste, for crawling to be worth the risk. Diarrhea can be deadly for infants in developing countries, and a study of Bangladeshi infants found that the incidence of diarrhea rose along with the proportion of babies who were crawling (but not yet walking). A baby who crawled was a baby more likely to get diarrhea. The traditional child-rearing wisdom of rural Turkey and New Guinea was well founded.

But these were not prohibitions confined to isolated, premodern societies. In the United States and Europe, crawling has only recently emerged from a cloud of deep suspicion. In fact, only a few decades before Gesell made crawling a requisite stage of locomotion, babies in America mostly *didn't* crawl.

The Western prejudice against crawling dates to at least the Middle Ages, when babies who crawled were called "kittens"— not because they were cute but because they were behaving like another species altogether. It was feared that crawling lowered infants from their godly purpose. This fear would persist: the Puritans in the New World believed the same thing.

These subhuman associations with crawling—squalidness, beastliness, godlessness—were so strong that the act was not suitable for polite discussion. In his 1690 child-rearing manifesto, John Locke reassured readers that "want of well-fashioned civility in the carriage . . . should be the parents' least care," a reference to crawling—but, as the historian Karin Calvert notes, even Locke, who was telling parents *not* to care about crawling, "could not bring himself to use the term." And Locke was far more progressive than his readership. For many of his readers, letting the child crawl like an animal was tempting fate. It was wiser to keep him erect, just in case.

To make sure he stayed erect, that baby was placed upright in wooden walkers, not unlike the plastic models you still see today. These did little for short-term safety—the premodern household was preposterously dangerous and a clever baby could maneuver a walker into an open hearth—but preventing crawling was about long-term safety: safety before God. Standing stools, which were often nothing more than a carved-out tree trunk, were more secure: they had the advantage of preventing much of anything at all. Some stools were so tightly configured that they *squeezed* the baby into a upright position. He had roughly the same chance of escaping as a Pringles chip does its can. But even outside of wooden walkers or standing stools, babies had difficulty going anywhere: for centuries, children grew up under heavy, bulky clothing that strangulated all movement. Even the lighter clothing was too long for a baby to get traction.

The Enlightenment encouraged a new acceptance of children as children: for the first time, many infants were let loose to explore on all fours. Decades later this was still a daring act, though. In an 1835 journal she kept of her newborn, the American mother Almira Phelps admitted that she'd planned to prevent her son from crawling—but she couldn't do it. Watching him, she'd decided that crawling was nature's way. Her acceptance of her baby as a baby was typical. In the nineteenth century, parents began to believe that babies would develop on their own, without adult guidance and without turning into quasi-human quadrupeds. They'd find their own way to walking.

This laissez-faire attitude had its limits: even at the turn of the twentieth century, every other child never crawled. In 1900, the psychologist August Trettien published a lengthy paper that attempted to systematize the "peculiar purposeless character of the movements of the human infant" but ended up being a catalogue of eccentricities. Like so:

> M. Harold went along like a frog. Laying palms of hands on the floor, then with a little jerk of his body would land on his knees. He would then begin again with his hands and repeat the same process. He progressed quite rapidly considering his method.

In Trettien's sample, only 42 percent of babies crawled on their hands and knees.

And yet a few decades later, crawling was cemented as a crucial, unskippable stage in motor development. In a mere half century, it had gone from being controversial to being mandatory. Suddenly, *it was always thus,* even though, of course, it had never been thus.

A century after Gesell enshrined crawling as a developmental milestone, it remains there, a bit tarnished but still bronzed. A parent today who wonders if his late-blooming child really needs to crawl may be told, by seemingly trustworthy sources, that crawling is indeed extremely important. "Skipping this milestone can also affect a child's ability to hold silverware or a pencil down the road," according to an article on crawling in *Parenting* magazine, "since the weight-bearing experience of crawling helps develop arches and stretch out ligaments in the wrist and hand that are needed for fine motor skills."

But at least these are relatively minor things: bad handwriting, poor silverware management. Or maybe not.

"What's so important about crawling?" an occupational therapist writes in a post on her hospital's blog:

> When a child begins crawling, this repetitious movement helps stimulate and organize neurons, allowing her brain to

control cognitive processes such as comprehension, concentration and memory.

These are no longer minor things! Plus, the therapist continues, crawling—not just any crawling, but symmetrical hands-and-knees crawling—develops hand-eye coordination and binocular vision. *Plus,* infants who don't crawl may miss the opportunity to integrate their "symmetric tonic neck reflex," and a poorly integrated STNR may result in learning disabilities and ADHD.

At this point, the aforementioned parent, who'd wondered if crawling was actually important, is pretty obviously flipping out.

If this parent searches further, he may find other articles arguing that crawling effectively knits the hemispheres of the brain together. There are EEG studies cited: the onset of crawling is shown to coincide with a spike in neural activity.

By now, the aforementioned parent is probably forcing his child to crawl through hoops.

And if he searches still further, he may discover the Institutes for the Achievement of Human Potentials, founded a half century ago by the physical therapist Glenn Doman and the psychologist Carl Delacato. At which point he flees the house.

Doman and Delacato believed that ontogeny recapitulates phylogeny—the century-old idea that as every individual organism develops, it proceeds through the evolutionary past of its species. The human brain, they believed, only reached its full modern potential when it was "properly organized." Each stage of motor development corresponded to neural development and each had its proper place, an idea that was Gesellian in spirit. But Doman and Delacato went further: they said that if a baby proceeded to a stage out of order, he should be immediately stopped. If he wasn't, his neurological system would be "disorganized" and his language development impaired—he wouldn't reach his full human potential. For children with developmental problems, Doman and Delacato prescribed "patterning," a physical therapy in which the child's limbs were moved for him as if he were crawling himself. By simulating crawling, the symmetrical movements were supposed to "reorganize" the damaged brain. Skipping crawling was the problem; simulating crawling was the solution.

These ideas weren't obscure. Doman and Delacato were featured in *Life* and *McCall's*. Satellite institutes of theirs opened around the world. Thousands of parents of brain-damaged children came to the institute, desperate for help.

But the biogenetic law on which Doman and Delacato based their theory—ontogeny recapitulates phylogeny—isn't a law. Biologists consider it completely specious. The idea of proper neurological "organization" isn't any more valid. The Yale psychologist Edward Zigler evaluated the Doman-Delacato therapy and published a paper with the anguished title, "A Plea to End the Use of the Patterning Treatment for Retarded Children." The American Academy of Pediatrics has repeatedly issued position statements against patterning. Nevertheless, the institute Doman and Delacato founded still exists. It still has offices around the world. And it still touts the neural importance of crawling.

With ominous warnings like these, it is no wonder that the baby boards of the Internet are crowded with parents worried about creeping and crawling: "Walking without crawling=learning disability??" a mother posts plaintively. These worries are diverse and diffuse: they expand to fill the available space. Some parents are concerned because their baby never crawled. Some are concerned because their baby crawled too much. Some are concerned because their baby crawled too little.

Arnold Gesell thought hands and knees crawling was important. "He wasn't really interested in crawling on hands and feet," says Karen Adolph, a psychologist at New York University. "He wasn't interested in pivoting. He didn't do anything about rolling. But everyone knows that babies do those things, too. He just didn't pull those out as being important milestones. It's arbitrary which things you highlight and make items on an assessment scale. The fact that Arnold Gesell was into certain skills—does that mean now we have to be into those skills? No, but we are."

No scientific study has ever linked not crawling to any negative outcomes. *None.* (Kids with developmental problems are more likely not to crawl. But that's a separate issue: they're more likely not to do all manner of things.) But the study of infancy is very neatly compartmentalized. The various spheres of knowledge— the neurological data, the ethnographic accounts, the historical documents, the psychological experiments—never have to acknowledge each other. An article about the crucial neural significance of crawling will never pause for an interdisciplinary mouthful of a disclaimer like, *But considering the vagaries of crawling as a milestone, and the long-standing historical and cultural bias against crawling, it seems unlikely that such critical neural connections could only be formed through crawling.*

Bizarrely, these beliefs about crawling persist even though many infants continue to not crawl. A study found that nearly *a fifth* of all British children never crawled on their hands and knees: they scooted around on their bottoms instead. Another 7 percent didn't scoot, or commando, or crawl—when the time came, they just stood up and toddled away.

Academics can be as provincial as parents: they often assume that some rough version of their research is how things should be. Read enough work on infant development and you begin to feel like you're back at the playground, talking to very helpful parents: everyone knows One Big Thing. Everyone rotates around their One Big Thing.

But what if thinking about One Big Thing is the exact wrong way to think about development?

16

Where Movement Comes From

After Gesell, the study of motor development entered a steep decline. This was partly because Gesell was so thorough. He'd taxonomized twenty-two separate stages of crawling. What was a new researcher supposed to achieve—the groundbreaking identification of a twenty-third stage?

For almost a half century after Gesell established his norms, the proper sequence and timing of development were treated as sacred. Any deviation was dangerous. New research was largely limited to creating more precise norms—to making the norms yet more normative. Everyone liked norms: they provided parents and pediatricians with guidelines and expectations. Their siren song was too bewitching to ignore. Better yet: no one wanted to ignore it.

"It seemed as though researchers knew everything they needed to know about motor development," the psychologist Esther Thelen would later write about this period. "It provided the universal, biological grounding for the more psychologically interesting aspects of early development—cognition, language, and social behavior."

Esther Thelen was an accidental psychologist: at home with two preschoolers, she took her first graduate course at the University of Missouri, she wrote, "as a way of expanding my interests beyond jello cubes and *Sesame Street*." She ended up, also by accident—"Well, what class was the bored housewife to take?"—in an animal behavior course, which then led her, *also* somewhat accidentally, to conduct research on wasps, specifically a species of parasitic wasp that conducted an elaborate antennae-rubbing, leg-waving courtship dance. These dances reminded her of the repetitive, rhythmic movements that all infants produce, for which no one had a good explanation, and she somehow convinced her advisors that these infantile tics would be an excellent topic for a dissertation. Her research impressed enough people that Missouri gave her a half-time position on the faculty and a small house next to the psychology building to use for her lab. The house had tile walls, drains, and cement floors. It turned out to be the old morgue.

Parasitic wasps, babies with tics, the morgue. There isn't much in this story so far that makes you think it'll end well.

During her dissertation research, Thelen concluded that the standard accounts of motor development were incomplete. She was intrigued, she wrote, by the "Case of the Disappearing Reflex. If you hold newborn infants upright by supporting them under the armpits, and then lower their feet to the table, they will take 'steps'—an alternating kind of march that is always quite surprising to see, because newborn infants normally are so uncoordinated." This behavior disappears after a couple of months; it was thought to be inhibited by subsequent neural development. But Thelen noticed that infants who no longer displayed the stepping reflex still kicked when they were lying down, and that their kicking looked a lot like stepping. She mapped it out: it *was* stepping—the movements were exactly the same. It didn't make sense for the reflex to be cortically inhibited while an infant was standing up but not while the infant was lying down. So there had to be another explanation—an explanation that wasn't purely neurological.

Thelen thought about other explanations for a while. Then she went out and "bought the largest fish tank we could afford." Her logic was deceptively simple: shortly after birth, babies gain a lot of weight, especially in their legs. This weight is fat, not muscle, so it would make it harder for infants to step when standing up. But it wouldn't make it harder for them to kick when lying down. Thus the fish tank: underwater the limbs wouldn't be heavier; they'd be buoyant. If the increased weight had prevented stepping, then reducing the weight should restore it. And it did. When Thelen's lab put infants in the fish tank—"It was a bit tricky explaining to parents why we wanted to hold their 2-month-old infants in torso-deep warm water in a large fish tank"—the stepping immediately reappeared. The stepping had been inhibited, after all—but by physical development, not neural development. It wasn't the cortex. It was the fat.

This was revolutionary. According to Gesell, the body was the puppet of the mind. But in Thelen's experiment, the puppet was talking back.

Since Gesell, motor development had been regarded as, well, *boring*. Remember how Thelen characterized the conventional wisdom: that motor development set the stage for *the more psychologically interesting* forms of development. For a long time, this was the damning consensus about motor development versus every other sort of development. It was interesting if you were a parent, of course—has any parent ever not felt a thrill when his child took her first step?—but it wasn't scientifically interesting. It wasn't mysterious or extraordinary. After Gesell laid down his stages, it was taught that infants learning to walk obey a set of neurological instructions. They just do what their brains tell them to do. Erased from their own development, babies became the passive subjects of their own movements. You couldn't see the personalities for the neurons.

From her humble fish-tank-in-a-morgue beginnings, Thelen would formulate a radically new theory of motor development. Hers told a very different story: that learning to move is among the most complex, ingenious things we ever do. It isn't boring. It isn't generic. It's wondrous.

Esther Thelen once asked the most basic question you can ask about how babies move. It's so basic a question it is almost meta: How can a learner *who does not know what there is to learn* manage to learn anyway?

Ever since Isaiah first struggled to sit up, it had been sort of embarrassingly obvious that he had no clue what he was supposed to be doing. He wasn't *trying* to do anything at all. I'd assumed learning to walk was a little like assembling a puzzle: you looked at the puzzle box, you looked at the puzzle pieces, you tried to make the two match. If your puzzle sucked, that was because puzzles were hard; you kept fixing the puzzle until it didn't suck anymore.

This analogy did not correctly describe Isaiah's development. It did not even approximate it. It wasn't that Isaiah was worse at putting together the puzzle than I expected. It was that he had no idea there was a puzzle involved. He didn't look at people walking and think, *Now that is a splendid idea.* It was more like he started to walk and then looked at people and discovered *they were doing the same thing he was.* My parents had a dog who'd occasionally wake you up barking in the middle of the night, and when you opened the door to see what the problem was, she'd look at you with an expression that said, *Really—you're up, too? What a crazy coincidence.*

Isaiah was a little like that.

The answer to Thelen's question, as her colleague Linda Smith put it, is that infants "can discover *both* the tasks to be learned and the solution to those tasks through exploration." It's *exploration* that's the key word here: each child discovers the facts of movement and the mechanics of movement in his own particular way, and these discoveries are all made through exploration. This explanation has the advantage of clearing up a lot of mystifying behavior. "Young mammals—including children—spend a lot of time in behavior with no apparent goal," Smith wrote. "They move, they jiggle, they run around, they bounce things and throw them, and generally abuse them in ways that seem, to mature minds, to have no good use." But it only seems that way. In fact, it is the very *best* use of an immature mind. Such screwing around, according to this way of thinking about development, is anything but screwing around. It's like a Buddhist koan theory of development: goals are achieved only through goallessness.

The actual theory Thelen developed, dynamic systems theory, has a couple of basic tenets: first, the puppet gets to talk back—the body itself plays a part in development. Thelen stressed that no component of motor development—the brain, the foot, the banana peel—is more important than any other. Movement is the confluence of these components. The brain can't translate an order into action all by itself.

Second, there's no "pure" behavior. Behavior only occurs in context: you put the kid in the fish tank, you get one type of behavior; you take the kid out, you get another. We act in accordance with our surroundings. In the moon's lowered gravity, as Thelen once pointed out, the astronauts immediately realized walking was made for the planet they'd left behind. So they started jumping instead.

These tenets seem logical enough, but once you piece their implications together, you end up with conclusions that are deeply counterintuitive. They suggest that learning to move isn't about obeying orders—*there isn't anything to obey.*

From the distance of adulthood, this seems sort of crazy. After all, we all end up walking. We must get there in the same way: we must be assembling the same damn puzzle. But when Thelen looked at motor development, she managed to stop seeing it as an adult would—someone who already knew what the infant would end up doing. She didn't just have to overcome the conventional wisdom; she also had to overcome the just-so logic of adulthood.

The trick is to look at motor development from the perspective of the infant—someone who has no idea what she is going to be doing. And once you're down there on the floor, you can see that, before walking emerges, there's something of a free-for-all going on. Infants aren't obeying. They're *discovering*. Motor development, far from being generic, is a creative act.

You can see this, in comic form, when you put a baby in a jolly jumper.

The jolly jumper, also known as a bumper jumper, is an infant harness attached to a spring and hung under a doorframe. When a baby sits in the harness, his feet brush the floor. If he pushes off it, he'll bounce—that's the point of the thing. But the baby, of course, has no idea what the point is. He has no idea he's even in a jolly jumper. So how does he find out how this thing works? As the psychologist Eugene Goldfield described the problem, "There are no instructions or models—the behavior of the system must be discovered." The problem of the jolly jumper is similar to that of reaching, or sitting, or walking—all tasks in which babies have to discover "the behavior of the system" within the constraints of their environment. It's just that with a jolly jumper, the environment is more narrowly defined: the harness, the spring, the floor.

To figure out how the baby figures it out, Goldfield calculated the "specific resonant frequency" of the jolly jumper, the frequency at which the jumper was most efficient—at which the infant got the most bounce. Then he ran an experiment—a jolly jumper experiment. At first, infants in the jumper stabbed experimentally at the ground with their feet, as if testing what would happen. Over the next few weeks, the infants, as if sorting through their experimental data, perfected a bouncing pattern that had nearly the same resonant frequency Goldfield had calculated. The babies discarded counterproductive bouncing patterns and kept the patterns that worked. They realized, for instance, that for maximum bouncing they should kick with both legs simultaneously—a movement they'd likely never made before. Over several weeks, the infants had puzzled out what the most optimal behavior in a jolly jumper would be—even though they hadn't had a clue about what the jumper was meant to do.

In order to get around in the world, or to get a jolly jumper to work, an infant can't rely on what he already knows: he has to discover how to solve each new problem. The world, in miniature, is a jolly jumper: a strange new toy with strange new rules.

Inventing Movement, One Baby at a Time

Let us pause to consider the scale of the task:

> Newborn infants face a profusion of challenges ahead that need to be tackled with an initially very limited repertoire of skills. Not in the least, they have to figure out that they have a body, what its characteristics are, and how to control its movements.

Given the magnitude of these challenges, infants have a very sensible strategy: to be completely clueless.

This is not slander. Isaiah learned to walk despite never having any idea he was learning to walk. His haphazard, accidental path toward locomotion is entirely typical of how babies learn to move. They are not efficient; they are not predictable; they are not in pursuit of any obvious objective. They are like the blindfolded child who manages to hit the piñata but first makes contact with most of the other partygoers. If you swing wildly enough times, you *will get the candy.*

Motor development is a few thousand wild swings of the bat. Watch the unsystematic, stuttering way infants move in the vague direction of bipedalism. They frog-leap forward to an advanced skill and then retreat to a stage they'd skipped. Rather than adopting a new skill and discarding the old, they flirt with new abilities for weeks, the way a cat toys with a mouse—picking it up, dropping it, picking it up again. They don't move upward in a steady Ascent-of-Man climb toward competency. They roll, they hitch, they scoot, they peg-leg crawl. They goose-step. They do the bunny-hop. They screw around.

Until recently, no one knew what to make of all this screwball behavior. For many years, any hesitant, nonlinear progress was dismissed as meaningless—static that obscured what was actually happening: the precise, orderly acquisition of motor milestones. Variation is the antithesis of the traditional, Gesellian view of motor development, which is why textbooks on the subject described variation as error—as babies doing it wrong. Some still do.

Some variation may just be sloppiness. (Babies are sloppy.) But a lot of it may be functional: playing around is good preparation for the messiness of the world. You'd think learning to walk would be like learning to play the piano: you work hard at a very specific, very focused skill. But a piano is always a piano: there are never eighty-nine keys or eighty-seven. The world around you, on the other hand, is constantly changing: walking is like having to *always* play a piano with eighty-nine or eighty-seven keys. In real life, pretty much the only time a person walks straight ahead on a predictable surface is during a DUI test. Knowing how to walk straight ahead on a predictable surface doesn't help if the living room floor is covered with Legos. Being able to adjust to the discrepancies of everyday life—the inconvenience of how things actually are—is the most important thing to get right.

This is simultaneously obvious and odd. Most of us instinctively assume that babies have mastered a motor skill when they settle on a single consistent version of it—when they stop screwing around. That's when the skill looks, well, skillful. But it turns out that babies who rely on a single gesture for, say, reaching are very bad at reaching. They only improve when they experiment with many different gestures. What we assume is mastery—refined, repeatable actions—is what holds infants back from attaining it.

Learning to move is an experiment run perpetually, each time tweaked. When puzzling out how to sit, for example, babies vary their posture in many minute ways, as if feeling out what's possible and what isn't. A baby who doesn't vary his posture will only learn how to sit a single way, which is to say that he won't learn how to sit very well.

He also won't be able to adjust to his rapidly changing body. This is crucial, since infancy is nothing but rapid change. In the movie *Big*—the story of a boy who wishes himself into an adult—Tom Hanks wakes up with a new grown-up body and no clue how to use it. But Tom Hanks had it easy: he only had to learn how once. In infancy, waking up with a new body happens over and over again: babies who haven't grown for a couple of weeks sometimes sprout up to one and a half centimeters in a single day. If babies were locked into a set of standardized movements—the way head-down theories assumed—they'd be helpless to cope with the peculiarities of their own bodies. And given the speed and scale of growth in infancy, their bodies are often really, really peculiar.

Variation isn't an error, in other words. It's adaptive. It may even be a marker of *typical* development. An analysis of infants with standard versus delayed development found that the latter moved in a much more repetitive and consistent way. They didn't vary their behavior; they did the same thing over and over again. The healthy infants, meanwhile, spent their time screwing around. Another study found that the absence of varied movements in the early months of life correlated with later developmental problems. Because of studies like these, assumptions about what's normal have been turned upside down: new therapies for infants with poor motor development focus not on showing babies "the right way" to move but on increasing the variety of experiences they encounter.

In the end, though, all variation is tightly constrained, hemmed in not just by gravity but by physiology. After rolling under the couch, or bottom-scooting across the kitchen, all infants at some point have the same epiphany: their bodies are made for walking.

No one knows just why babies walk when they do. Different theories emphasize different factors: new neural networks might make babies faster at responding to the challenges of movement; experience might train them how to deal with the tricky problems of balance and banana peels; a less babylike body might be stronger and more stable. There's no reason these factors can't all be right. What seems clear is that walking—the fundamental human trait—isn't built-in, locked away in some inviolable part of ourselves and our genome. It's discovered anew by every infant.

Walking: Why Bother?

After babies hit upon the ineluctable fact of bipedalism, they stand up, awkwardly, and step forward. And then they fall. It's no wonder. The mechanics of walking, when you think about it, come down to moving forward *while on only one leg*. It should be a circus trick.

The average toddler falls seventeen times an hour. But every hour he takes an average of more than 2,300 steps. If he took all those steps in the same direction, which no self-respecting toddler would ever do, they'd add up to the length of more than seven football fields. These are astonishing numbers, because when babies begin to walk, they have almost every reason to do it very badly and go nowhere at all.

These reasons begin with balance. If 90 percent of success is just showing up, 90 percent of walking is just staying upright. Here's the problem: as a matter of physics, staying upright is more difficult for babies than it is for you and me. The shorter you are, the faster you sway; the faster you sway, the less time you have to stop yourself before you sway too far. Infants learn how to balance when balancing is tougher than it will ever be again. It's like teaching a child to play the violin before the onset of hand-eye coordination. It's not a good idea.

Motor scientists have special language for the amount you can sway and not fall over—they call it "the region of permissible postural sway." If you stray outside that region without sufficient strength to right yourself, and almost no toddler has that sort of strength, you tip. Once upright, we are in continual danger of straying outside that zone. Every physical action—not just big movements like bending over, but small ones like idly thumbing your earlobe—alters a person's center of gravity. Adults unconsciously compensate for this: our back and abdominal muscles activate *before* we reach for something. Even when we think we're still, our bodies remain in very slight motion, so slight we're oblivious to it. To be upright is to be off balance.

This is part of the problem. The other part is that infants, while learning *how* to balance, also have to learn to throw themselves *off* balance: it is the only way they will move forward. Babies on the brink of walking sometimes appear paralyzed by this insight; they stand in place, dumbly, for long stretches. To break the deadlock, different infants settle on different strategies: some twist, rotating each leg forward in turn; others inch forward, moving almost imperceptibly; the most impetuous flat-out fall, tipping forward, righting themselves, tipping forward again. As new walkers, they live in a contradiction. Their movements are guided by the imperatives of balance, but these imperatives *conflict*: you must always be balanced except when you must be off balance.

Nevertheless, they manage. With their flat feet splayed outward like ducks, toddlers wobble along, somehow staying upright, despite the best attempts of their body to topple over. Their feet are widely set and their arms high and guarded, as if they sense that gravity is trying to push them down. They are solving the problem of balance but it comes at a cost: dignity. Watching Isaiah was not all that different from watching a Buster Keaton short.

A few days after Isaiah took his first steps, I looked down and wondered why he was even bothering. A few days before, he'd been very good at crawling: fast and efficient and in close proximity to various foodstuffs. If he wanted a toy, he went and got the toy. But he was a disaster at walking: if he wanted a toy, a minute later he still wanted the toy. He was stuck in the middle of the floor, entangled in his appendages.

I knew walking was worth the effort. But how did *he* know that? The feedback Isaiah was receiving was uniformly negative. When I receive uniform negative feedback from an activity, *I stop doing it*. Why wasn't he stopping?

Karen Adolph, the psychologist at New York University, has seen many babies who are terrible at walking. For almost a century, psychologists like her who study motor development have measured how and in what ways babies are terrible at walking. To do this, over the years they have tracked:

the scratches in the floor left by hobnailed boots, chalk impressions of infants' feet on black photographic paper, and series of footprints made from olive oil sprinkled with graphite, ink-coated corn plasters, or from walking through talcum powder. Perhaps most ingenious was the technique devised . . . for capturing the placement of infants' steps. Babies walked over an array of tiny, black, rubber cones and white evaporated milk, sandwiched between two glass plates. As their feet contacted the glass surface, the cones deformed and displaced the milk revealing a real-time trace of babies' footprints.

With the advent of highly precise recording equipment, tracking movement is far easier. The ingenuity is in experiment design. Many psychologists studying motor development construct experiments that are exaggerated versions of everyday life— exaggerated to make sure the infants have never encountered anything like it before. These experiments—Adolph has designed a study involving Teflon floors—look outlandish to adults. But to babies, such scenarios—highly artificial, fantastical scenarios— look completely realistic. They're just another weird feature of the world: yesterday, escalators; today, Teflon floors.

Adolph is a genius at designing these experiments (Teflon-soled *shoes*). She's designed experiments with walkways that suddenly morph into huge foam pits. Reading her work, you imagine her laboratory looking like the set of a Nickelodeon after-school special.

As it happens, her lab had recently conducted an experiment that seemed to answer my question about walking—*Why bother?* The answer, although Adolph was too kind to say so, is that I was wrong about everything I thought I was seeing.

To begin with, crawling may seem fast and efficient, but it can only be *so* fast and efficient: a very good crawler is slower than a very bad walker, even in the first week of walking. At a year old, some babies walk and some crawl—the walkers are novices and the crawlers experts. Nevertheless, the novice walkers travel three times as far and take twice as many steps as the experienced crawlers. "Twelve-month-old walkers fall 32 times an hour and 12-month-old crawlers fall 17 times an hour on average," Adolph says. "But the walkers are moving more." So much more, in fact, that both groups take the same number of steps per fall. Babies do not pay, as Adolph puts it, "a greater penalty" for starting to walk.

But they do receive vastly greater rewards. Any infant can do the math.

But I'd screwed up the math: after months of watching Isaiah crawl, I no longer noticed when he fell while crawling (and crawlers fall way more than you'd expect); I only noticed when he fell while walking. These falls seemed dramatically worse to me—they were from higher up. But they didn't seem worse to him, maybe because the experience of walking was exhilarating. He was moving much farther, but I only noticed when he stopped moving—when he hit the floor. Even experienced walkers fall a lot. But babies barely notice these falls. Only their parents do.

In my defense, I was less hysterical about it than, you know, *some* parents.

For a few centuries, up through the 1800s, infants learning to walk often wore a helmet that was called, wonderfully, *a pudding*. These tended to be the children of well-off families, the sort who had nurses and nannies, and the infants were well protected by their social status and their helmets. The pudding was a very thick strip of padding, wrapped tightly around the forehead and tied under the chin. It inspired a splendid nickname for toddlers: *puddin' head*. But for some parents even the pudding was insufficient protection. Some children were attached to leading strings, fabric leashes designed to prevent them from falling: if they started to topple, the string would yank them erect. Imagine a puppet with an overprotective puppeteer.

But beware the child-proofed child. The leading strings created their own problems: when a child didn't want to walk, some nurses were said to *drag* him by the leading strings rather than carry him.

On the other hand, at least the kid didn't fall.

Getting Better All the Time

In *The Principles of Psychology,* William James, doing his best evil genius impersonation, wondered if:

> a baby were kept from getting on his feet for two or three weeks after the first impulse to walk had shown itself in him—a small blister on each sole would do the business—he might then be expected to walk about as well, through the mere ripening of his nerve-centres, as if the ordinary process of "learning" had been allowed to occur during all the blistered time.

"It is to be hoped," James concluded, "that some scientific widower, left alone with his offspring at the critical moment, may ere long test this suggestion on the living subject."

The inspiration for this suggestion was the baby swallow: James had read that swallows caged from infancy will immediately take flight when their cage is opened. The swallows didn't learn to fly; they knew how. Clearly, and you can almost hear James muttering this to himself, caging human infants was not practical, but *a small blister on each sole would do the business* . . .

The idea of walking as instinct dominated psychology for decades, even though no scientific widower ever emerged with proof of it. But if infants did *mature* into walking, the way James imagined—if walking clicked at a particular age—they'd have a stock set of instinctive movements. Instead, infants move in a wild and bewildering number of ways—a stock set of movements is precisely what they don't have.

Instead of age, experience is the best predictor of success at crawling or walking. In experiments that required zero blisters, Adolph has confirmed this finding over and over again. Overall age is actually a poor predictor of success. How long an infant spends crawling or walking—how long they have had to practice—is far better.

Think of it this way. There's no *abstract* walking—you can't walk in your head. The environment, no less than the brain and the rest of the body, dictates how walking develops. Walking doesn't count unless you're walking *somewhere*.

What babies learn from practice is how to solve the problem of *somewhereness*. This is where things get screwy. For a new walker, the most obvious moment to learn would be when something goes wrong, and the most obvious evidence for something going wrong would be a fall. But babies show no sign of making this connection. A baby who topples over headfirst is not any better at walking afterward. He isn't any better at not falling, either; he isn't even more cautious. In an Adolph "foam pit" experiment—in which a padded floor suddenly turned into a foam pit—infants who were experienced walkers repeatedly tumbled into the pit despite the clear presence of "bumpy foam blocks" and "blinking lights." It's not that they didn't notice the warning signs: "One 15-month-old stopped to touch and admire the blinking lights, but he resumed walking and fell into the foam pit like the rest of the children." Nearly a quarter of all infants in the experiment fell into the pit in sixteen straight trials.

In the context of the foam pit experiment—in which the blinking lights and bumpy blocks are all but shouting at the infant to *pay attention*—an inability to learn from falling makes no sense whatsoever. But as a strategy for the more mundane falls of infancy, it turns out to be perfectly logical. The average infant falls as many as ninety times a day, but the vast majority of these falls aren't caused by foam pits, or wet floors, or uneven sidewalks. They're caused by the infant himself—babies, like drunks, fall down all on their own. This is why the learning curve for falls is so excruciatingly graduated: for a baby, the typical fall means nothing at all. When I trip, I turn around to see what I tripped on.

But when babies trip, there's no point in them looking back—they almost certainly tripped on nothing at all. Infants learn very little from falling, as Adolph has written, "because most of the time there is nothing to learn." Before a baby can learn from a fall, the fall itself has to *mean* something.

But when infants link actions with consequences, they learn very rapidly, devising impromptu solutions to new problems. In a couple of experiments conducted by Adolph, this spontaneous creativity is on full display. The experiments examined how sixteen-month-old infants, all capable walkers, dealt with crossing a bridge that stretched over a short gap. When the bridge was wider, the infants scampered across without using the available handrail; when the bridge was narrower, they clutched the rail. In both instances, they hardly ever fell: they made snap decisions about what to do correctly. In a follow-up study, Adolph and her colleagues used different handrails, some solid, some wobbly:

> To our surprise, infants preferred a wobbly handrail to no rail at all even though the wobbly handrails were flimsy and could not support their full weight. They crossed narrow bridges more often with the wobbly handrail than infants facing the same bridge widths with no handrail in the previous study, and their attempts were largely successful. How did infants manage to outwit three seasoned experimenters' best-laid plans? Infants devised clever, new solutions for crossing the narrow bridges by exploiting the wobbly, deformable properties of the handrails.

They did the "hunchback," facing the handrail and draping their body over it until the rail was as low and stable as possible. They "windsurfed," facing the handrail and holding on while leaning backward for support and sliding sideways across the bridge. They "mountain climbed" across the bridge; they "snowshoed" it. They acted like they'd had a few too many, sliding forward with all their weight against the rail, like someone descending a stairway in the wee hours. They were flexible beyond expectations.

Confronted with a novel challenge—so novel the experimenters thought it *too* challenging—the infants devised an array of clever little tricks. And their tricks worked! In these bursts of ingenuity, you can glimpse how babies go from slugs to bipeds: how the challenge of new tasks—every day, in every new setting—pushes them to elaborate and refine their abilities.

Rather than matching a stock movement to a stock situation, the life of an infant requires having to solve a new problem, and then another, and then another. It's a tremendously inefficient way to stockpile answers, but the point of it isn't acquiring knowledge. The point is acquiring *how* to acquire knowledge. Infants aren't memorizing answers; they're learning *to come up with* answers, at the speed of seven football fields an hour.

Once an infant is any good at crawling, or walking, or any mode of locomotion, she's a keen judge of her limitations. In experiments in which a baby who's a skilled crawler or walker encounters a challenging obstacle, she will scrutinize it before proceeding. Sometimes she will flatly refuse to go on. But a baby who has just acquired a new motor skill is a terrible judge of her limitations. Looking at the world from a new posture, infants seem to have no recollection of ever having interacted with it before.

This is weird enough to require some elaboration.

At nine months, most infants are novices at crawling but have mastered sitting. If they sit on one side of a gap with a toy on the other side, they know exactly how far they can reach safely. But if they are in a crawling posture—their new skill—they have absolutely no idea how far they can reach: they invariably reach too far and topple into the gap. If they try again, even if they try again immediately after falling, they will fall again. They will fall again even when the gap is absurdly wide, way too wide to reach across—when it isn't even a judgment call.

A few months later, these same infants are novices at walking but have mastered crawling. If crawling, they are now very accurate at calibrating, say, the risk of sliding down slopes. They know when they can make it and when they can't. But if walking, these same infants are horrible at assessing the very same risks—happily and without hesitation, they try to walk down slopes they would never have tried to crawl down. Placed in different postures, they behave as if they were different babies.

It gets weirder. Babies not only don't retain knowledge when they shift between postures, they don't learn any faster in the new posture. Each posture is so different, and their bodies have changed so much, that earlier experience is essentially worthless. The lessons they'd painstakingly learned while sitting don't apply to crawling.

Take cruising. You'd assume that cruising prepares infants for walking. After all, cruising very nearly *is* walking. But in experiments, infants cruising on a handrail will confidently step out onto nothing at all. They're fooled by the support under their hands. If the handrail has a sudden break, they find a way around it. But if the floor has a sudden break, they are totally oblivious to it. They're paying attention to the wrong thing. We know what stage comes next, so we assume cruising teaches babies about the ground. But babies have no clue what comes next. They don't know what to focus on. Which is why with every new posture, infants require some twenty weeks of practice before they stop committing these unforced errors.

We think babies learn to walk by progressing through stages, more or less like Gesell said: sitting, then crawling, then cruising, then walking. The knowledge they gain on their bellies, and then on their hands and knees, builds on itself, just as an intermediate Spanish course builds on Spanish 101. But if crawling isn't preparation for cruising, and cruising isn't preparation for walking, then the development of locomotion isn't at all like someone progressing to fluency in Spanish. It's like someone taking Spanish 101, and then the history of Japanese puppetry, and then multivariable calculus. The content is completely unrelated.

"Is This Child 'Normal' or Shall I Take Him to the Clinic?"

That's from the foreword to a book by the president of the British Psychological Society, published in 1956, and it neatly summed up the concerns the book was meant to address. Its splendid title: *The Normal Child and Some of His Abnormalities.*

A half century later, the language for talking about these things hasn't gotten much more precise.

After we moved across the country when Isaiah was a toddler, we had to complete a one-page questionnaire for his new day care. The form was constructed like a Mad Libs game, with blanks for us to finish the sentences, and about halfway through it, we hit these:

Developmentally, my child is _____.

My child walked at _____.

Isaiah at this point was in his mid-twos. He'd already run the terrifying gauntlet of early milestones. I had thought we were through with this.

We weren't. At an interview a few months later for a preschool—long story, honest—the admissions director asked us, very earnestly, "And do you remember when he started to walk?" We tried to remember. "And is he meeting all his developmental milestones—running, climbing?" *Climbing?* Climbing what? We nodded numbly.

A wave of new motor development research may have over-turned the paradigm on which developmental milestones were founded—the omnipotent power of neural maturation—but the milestones themselves still stand. They don't even wobble much. Developmental psychologists have pretty much discarded the concept of developmental milestones, but developmental milestones remain the only thing that most people know about developmental psychology.

No parent would have much reason to suspect that milestones are suspect. They are still prominently featured in child development textbooks and pediatricians' offices. They're around mostly because they never went away. No one advocates for milestones; they aren't endorsed by the relevant medical journals or committees. Nonetheless, they continue to be the most common tool pediatricians use to track developmental progress.

The vast majority of milestone lists—from standard pediatric references to sidebars in baby manuals—use the median age. A few lists provide a range of ages or some broad parameters. But the median percentile is what gets the major billing: unequivocal, memorable, it catches the eye of the doctor and the parent. It sticks with you.

Of course, half of all children will always be beneath the median—otherwise it wouldn't represent the fiftieth percentile. "In other words," as a pediatrician has written, "after reviewing 50th percentile milestone information, as many as half of parents could conclude that their child is possibly 'delayed.'" Instead of medians, it would make far more sense, in almost all circumstances, to talk about the tenth to ninetieth percentiles—the vast age range of normal variation.

Even these broad parameters risk being too narrow, though. In a recent study, forty-five infants were assessed monthly on their motor skills until they were eighteen months old. During this time, their percentile rankings for motor skills rose and fell wildly, for no apparent reason. Sometimes the babies had small flurries of accomplishment; sometimes they fell into a rut. Nearly a third of them ranked below the tenth percentile at least once—the red line for "at risk" development. But every one of them ultimately turned out just fine. None actually warranted the "at risk" label.

Time and again, studies have failed to predict future problems from motor development measures, and once you grasp just how many infants *appear* at risk, you realize why such studies keep failing: at any random point in time, a perfectly healthy baby is indistinguishable from an infant with developmental problems. By design, developmental screening is guaranteed to return scores of false positives.

Given how crucial early intervention can be, doctors and therapists will never stop screening for developmental problems, nor should they. But since it is so tricky to identify delays, they might never get much better at it, either. Medically, we may always have to trade a preschool's worth of false positives (needless anxiety about developmental problems) for a few accurate diagnoses (actual developmental problems). That's probably a trade worth making: for children with serious problems, early therapy can be extraordinarily beneficial. But the rhetoric of milestones doesn't imply that doctors and therapists have to make this sort of trade. It implies a precision that's the very opposite of the reality.

Even if screening can't be improved, the language for talking about screening certainly can be. At the very least, the word "normal" should be scratched from the developmental vocabulary: what's typical in infancy is *variation*. Rather than a long list of milestones, parents would sleep better with fewer but more relevant guidelines. We'd all do better to acknowledge just how unstructured infancy actually is. Later is just fine and earlier isn't any better.

Okay, but still, really: *Is earlier better?*

This is the question many parents will ask themselves at some point, possibly while looking down at their sixteen-month-old still on her hands and knees. It's all well and good, these parents might say, that infants develop motor skills at radically different times—that a baby who learns how to walk well after the age norm is still perfectly healthy. But what about being *better* than normal? Are babies who develop motor skills earlier better off than those who develop them later? Will they be smarter, more agile, more likely to succeed?

It's easy to find studies that make these claims. A recent University of Virginia study found that fine motor skills effectively predict academic success. Studies relying on extensive data from Finland have found correlations between the onset age of standing unsupported and muscle strength and fitness some thirty *years* later. Other studies, using some of the same Finnish data, found that the sooner infants stood, the higher their childhood IQ scores; by adulthood, these infants were more likely to have attained higher levels of education and to have higher reading comprehension skills.

It's also easy to find stories about infancy that repeat these claims. And reduced to a few stark conclusions, the studies look damning. Forwarded around the Internet, repeated in blogs and reposted on parenting forums, they ossify into conventional wisdom: earlier *is* better.

But there's less here than meets the eye. If you think I'm the only person saying so, that last study—the one linking early standing with later cognitive abilities—appeared in a journal alongside a highly critical editorial titled "A Case of Less Than Meets the Eye." From the editorial: "Perhaps our most pressing concern is that the readers of this journal not too readily accept the authors' interpretation that age of standing is related to, and a predictor of, IQ at age 8 years (and perhaps later as well)." Which was pretty much the whole point of the original article.

At the risk of getting lost in the weeds of experimental design, there are countless reasons to be cautious about matching motor skills with future cognitive abilities. In almost all large long-term studies, for instance, the age of skill attainment is rarely observed by the researchers themselves. It is reported, sometimes long afterward, by parents, who predictably report earlier ages than the actual onset age. Plus, correlations like those in the studies above are typically good for a single arbitrary skill (like standing). But a random correlation with a single skill doesn't mean much. What would be meaningful would be an entire *sequence* of motor skills: they should all correlate.

The thicket gets thornier. Is the parent's definition of a given motor skill the same as the researcher's—do they both define "walking" in the same way? Do they define it at two steps or twenty steps? Onset ages for a skill appear objective, but they can be highly subjective. Pinpointing just when a skill emerges is as much art as science: new motor skills frequently overlap with old skills, sometimes for weeks or months at a time. A baby who started walking yesterday may continue to crawl for the next month. What do you pick as the "real" onset age of walking, the first steps or the last day of crawling?

And what if the measurement of a motor skill measures more than just the motor skill? The study that found a correlation between fine motor skills and later academic success measured fine motor skills by asking children to copy a shape and draw a person. That's an exercise that draws on all sorts of different abilities—attention span and memory, for starters—that aren't motor abilities. But because it was called a test of fine motor skills, by the time it appeared in a discussion on my local neighborhood parent listserv, after it was highlighted in an article about preschools, the study was transformed—*poof*—into scientific proof that early fine motor skills are tied to academic achievement.

The people who spend the most time studying motor development know all about these variables, and not surprisingly, they are the people least likely to make grand proclamations about the future predictive value of motor development for anything. "If my baby started walking at 16 months, would I be like, *Nuts,* we're not going to Princeton?" Adolph asks rhetorically. "No."

That said, she adds, there is surely some *short-term* cognitive gain. This is a very different question than that of long-term gain, and a much easier question to answer. Take a six-month-old who's crawling and a twelve-month-old who isn't. "The six-month-old who's moving is getting information about the world that the 12-month-old can't get yet," Adolph says. An infant who can crawl or reach or stand has access to information he didn't have before. But this period of development has—let's all say this together—nothing to do with college admissions. There are no detailed, sophisticated, long-term studies that link early motor milestones with future achievement. There's no scientific reason to think that earlier is better.

There are cultural reasons, though. When Jean Piaget lectured in the United States, he was frequently asked whether the rate at which children attained his cognitive stages could be accelerated—in other words, whether you could train your child to be "ahead" of other children. Piaget was bewildered by the question. In his view of development, being "ahead" or "behind" anyone else was meaningless. But he got the question often enough that he came to associate it with a particular worldview: he called it "the American Question."

17

Just Your Normal
Milestone-Meeting,
Spear-Throwing Infant

In the late 1950s, a series of unexpected findings about infant development emerged out of sub-Saharan Africa. In test after test, African infants displayed far better motor skills at far younger ages than American or European babies. All these studies suggested a single sensational conclusion: African babies were *innately* advanced. They were physically superior specimens.

The source of the earliest studies was the French physician Marcelle Geber in Uganda, who reported some astonishing results: the Ugandan infants were at least a couple of months and sometimes as much as four months more advanced than their American counterparts. Ugandan babies sat up at four months, American infants not until six months. Other studies followed Geber's and they tended to validate her findings, especially for sitting, standing, and walking. A variety of vague race-based conclusions followed: Africans—and most studies quickly extrapolated from the sample to the continent—were, even as newborns, exceptional. And such differences, so early on, could only be explained by core genetic differences.

This logic sounds specious and it was. But here's the twist:

there is good evidence for "African infant precocity," as it was called at the time. It's just that the precocity isn't in the genes. It's in the parenting.

In the early 1970s, the psychologist Charles Super conducted a detailed, groundbreaking study of development in a Kipsigis community in Kenya. Over a three-year period, every infant born in the village had his motor skills tested monthly. The data was interwoven with ethnographic observations—what the daily routines and handling of the infants were like. This was the sort of multidimensional data—not just what the motor skills were, but how they developed—that had been missing. And it turned out that Kipsigis infants really were more advanced than their American counterparts at sitting, standing, and walking. They typically hit these milestones about a month sooner, and after only being in the world for a few months, being ahead by a month is way out in front.

But the Kipsigis were more delayed than American babies at "prone position" behaviors like lifting their heads, crawling, and turning over. "There is no generalized precocity, as one is led to expect by earlier reports," Super wrote. A genetic explanation wouldn't suffice.

The Kipsigis infants were sometimes ahead of their peers, Super realized, because they were doing advanced coursework in some subjects: almost all the mothers in the community taught their children how to sit, stand, and walk. At first, the teaching was incidental, part of the social fabric of village life. Shortly after birth, Kipsigis babies were placed in a sitting position, often in someone's lap, 60 percent of the time, far more than American infants. Then, at five months, the Kipsigis infants were trained to sit on their own: they were placed in a dug-out hole, which stressed and strengthened the muscles sitting requires. Shortly thereafter, they sat up on their own. Standing and walking were encouraged in similar ways. But because they spent so much time sitting and standing, Kipsigis infants spent less time on the ground than American infants. That's why they were worse at "prone position" behaviors—they hadn't developed *that* musculature.

The way the Kipsigis drill motor skills into their babies is typical of the region. In a survey of a dozen ethnic groups in East Africa, Super found that every group taught infants to sit and walk. He also found a group, the Teso people of Uganda, who, unlike all the rest, taught their infants to crawl. The result? Teso infants crawled earlier than the Kipsigis *and* the American infants.

Motor development wasn't a genetic inheritance set in stone. It was putty.

The Super study showed that motor skills, even very early in infancy, were highly malleable. If you really cared about sitting, you could get your baby to sit sooner. The same went for standing, crawling, walking: drill, baby, drill.

But these skills develop at different times not just because of explicit training but because of everyday routine. We obviously know that child-care routines vary—from person to person, from culture to culture. But it is breathtaking to see just how *much* they vary, especially for the most basic, mundane tasks. Take bathing:

In Mali, caregivers lift a young infant by gripping the baby under an armpit with one hand. They bathe infants propped in a sitting position in a bowl of water, scrub them vigorously with rough sponges, and shake them dry.

One wonders: *And I put an infant tub on my baby registry because why?* Over and over again, cross-cultural research on infancy teaches the exact same lesson: infants can tolerate—and thrive under—care that most any Western parent would assume would end very badly. And very different models of care mold babies who can do very different things.

It is the most basic care—the routines you are least likely to notice—that matters the most. Infants in most developed countries spend their days horizontally. They're like the sunny-side down eggs in *Bread and Jam for Frances*: they just lie on their stomachs and wait. Infants in most traditional cultures spend their days vertically, a position that promotes a wholly different pattern of muscle growth. When infants in these cultures are given motor development tests, they score highly in skills related to vertical postures but flunk those related to horizontal ones—like the Kipsigis did.

How a baby is allowed to move profoundly affects his motor development. There are a couple of dramatic examples of this— the first exotic, the second domestic.

The first is taken from some impoverished valleys in northern China, where infants a few weeks old were traditionally encased, flat on their backs, in small sandbags. The bag was babysitter and diaper in one: it restrained the child and absorbed his waste. In an environment with little water and copious sand, the sandbags were, in a sense, an ingenious child-care strategy. But in another sense, not so much: the babies were rarely removed from the sandbags; they spent more than sixteen hours a day in them (they had less time free than infants who were swaddled). And at almost a year of age, fully a third of the babies were not able to sit up.

The second example is more mundane—it comes from any American crib. In the early 1990s, following studies that found that infants who slept on their stomachs—then the vast majority— were at higher risk for sudden infant death syndrome (SIDS), the American Academy of Pediatrics formally recommended that babies be put to bed on their backs. Twenty years and a highly successful public health campaign later, only a tiny percentage of infants still sleep on their stomachs. The rate of SIDS deaths has been cut in half. But the AAP's recommendation had a side effect: American babies today are significantly slower to roll over, sit up, crawl, and stand than they were just a couple of decades ago. Why? They no longer get the routine upper-body exercise they once got from lying on their bellies.

So: if you slow infants down, they slow down; if you speed them up, they speed up. The question arises: How fast can you speed them up? What *can't* you get infants to do?

You may think you know the answer to this question. You are wrong.

The children of the Aka, the hunter-gatherer people of central Africa, accomplish tasks in infancy that would be inconceivable in America. Not just inconceivable—if you saw an infant trying to do these things, you would sprint across a busy interstate to stop him. The anthropologist Barry Hewlett, who lived with the Aka, said:

> I was rather surprised to find parents teaching their eight-to-twelve-month-old infants how to use small pointed digging sticks, throw small spears, use miniature axes with sharp metal blades, and carry small baskets.

Outside of carrying small baskets—something of a let-down after the spears and axes—teaching infants in the developed world to do any of these things would be considered a really bad idea. A criminally bad idea even—not criminally as in really bad, criminally as in *prosecutable*.

But within Aka society, it makes a surprising amount of sense: the Aka treat infants as if they are contributing, autonomous members of the community, and at a remarkably early age Aka children become exactly that. In a subsistence culture like theirs, in which survival itself is a struggle, these skills are essential. In such an environment, it ultimately may be safer to teach infants how to use an axe than to *not* teach them.

When taught, they do unexpectedly well. Aka infants handle sharp metal objects without slicing open everyone in the village; or, put another way, they successfully respond to the expectations others have for them. And Aka infants aren't an anomaly: among the Efe, who live in the Congolese rain forest, babies are said to "routinely use machetes safely." The Fore people of New Guinea allow infants to use knives and handle fire from when they start to walk.

None of this is easy to believe. When my father saw a photograph of an Efe infant using a machete, he said, "That's got to be a hoax, right?" My father was afraid I was being played, and a not small part of myself, as I am writing this, thinks the same thing: this *has* to be a hoax. Right? Reading the cross-cultural literature on infancy, this sort of skepticism rises up repeatedly: *There's no possible way a baby can do that,* you think. In this sense, the Aka feel like a thought experiment, explicitly constructed to upend the Western parent's idea of what's possible.

Our assumptions about what babies can and can't do are so deep and unconscious they don't even out themselves as assumptions.

The Rest of the World Thinks You're a Horrible Parent

The phrase "African infant precocity" suggests just how ingrained milestones had become a few decades after they were first established. It was taken for granted that the milestones of Western psychologists like Gesell were the true milestones—objective readings of how infants develop. When studies appeared of African infants achieving these milestones at earlier ages, the Americans weren't reinterpreted as delayed. They couldn't be. They were the objective measure! The African infants, therefore, had to be precocious.

The new trove of cross-cultural research undermined the statistical truths Gesell had spent his life nailing down. Gesell and many other psychologists had tried to prove, without ever actually saying so, that development existed outside culture—that it was autonomous. But these ethnographic accounts revealed not just that other cultures had radically different patterns of child development, but that each culture also had its own set of norms—each had an internal sense of what babies should be doing when. In other words, each culture seemed to produce, *de novo*, its own Arnold Gesell.

In a seminal study, a group of English and Jamaican mothers, all living in the same neighborhood in England, were asked to predict when their babies would hit certain motor milestones. Each group's predictions about their own children were largely accurate—even though the predictions of each differed by as much as several months. "There seems to be a culturally-specific match between maternal expectation and the actual ages at which major motor milestones are achieved," the authors of the study concluded.

And like Gesell, each culture has a refined sense of *why* these motor skills develop when they develop. The Kung, for example, like many societies that actively encourage motor development, believe that if children are not taught to walk they will never walk. They have no faith that the child will acquire these behaviors on her own.

From our neural-biased perspective, this attitude seems stubbornly dense. If children needed to be taught how to walk, wouldn't there be more people in the world who didn't know how to walk?

The Kung belief is what academics call a "folk theory"—the collective explanation a community has for a phenomenon. In cultures where motor skills are actively shaped, parents spend a lot of time thinking about the process of motor development, and a lot of folk theories emerge to explain that process. Western parents have lots of folk theories not for motor development but for what *we* think about: cognitive development, attachment, discipline. Motor skills are regarded as well-ordered neurological developments, and since we don't think parents can shape motor skills, there's little point in devising folk theories to explain them. Of course, our story about motor development is a folk theory, too, but we don't call it that; we call it science. For us, any theories about motor development that aren't "scientific" are de facto weird. And the Kung's theory of motor development has none of the trappings of science. It's automatically very weird.

The disjunct between what cultures assume, deep down, about development is apparent in a series of studies by the psychologist Heidi Keller. She had Nso mothers in Cameroon comment on the parenting of German mothers and vice versa, with each group watching footage of the other playing with their babies.

Things did not go smoothly.

The Nso mothers were deeply critical of how the German mothers handled their babies, beginning with something as basic as putting the child down on his back—a posture that would injure the spinal cord and cause long-term developmental problems, the Nso explained. The Nso train their infants to develop motor skills. "Lifting the baby up and down makes the baby grow well, faster and strong," they told Keller. When the Nso saw the Germans not doing this, they were distressed. The German mothers were setting back their children's motor development, the Nso mothers said. Watching the German mothers must have been an agonizing experience for the Nso: these inexplicable foreign women were doing nearly everything wrong. "The Nso even suspected," Keller noted, "that it may be forbidden in Germany for mothers to hold babies close to their bodies: 'They handle them as if they are not their babies, as if they belong to somebody else or as if they are babysitters.'"

Their verdict was unsparing: "'The Germans show a very bad example of child care.'"

The German mothers were as disturbed by the Nso as the Nso were by them. The Nso put their babies in bowls to practice sitting; they drill for walking, using bars for balance. When the German mothers saw how the Nso handled their children, they were shocked by how rough they were. An infant's body can't tolerate such stress, they said. But even though the German mothers were very much aware of the handling—they were more aware of it than anything else—they never linked it to the effect it had on the infants. "Although they acknowledge the accelerated development of the Nso babies—some even question their age of 3 months," Keller wrote, "they do not relate it to the motor practice of Nso mothers." After all, according to the folk theory of the German mothers, the infant is supposed to lie on his belly *and wait*. What's handling—what's parenting—got to do with it?

For the Nso, it matters if motor development is faster, because the sooner the child sits up, the sooner he can participate in the life of the community; the sooner he can walk, the sooner he can start to help, or at least be less of a burden. It's a connection almost completely absent from Western conceptions of child rearing, in which motor development is processed through inanimate objects—we talk about whether a child is coordinated enough to thread blocks onto a spindle. For Western parents like the German mothers, motor development is a means to cognitive development, not civic participation.

The Nso wanted to accelerate motor development. The Germans wanted to accelerate cognitive development. Trapped in their opposing philosophies of child rearing, each community was bewildered by the inexplicable practices of the other. Not just bewildered—they were *angered* by them. They both looked at the tape and saw parenting so bad it was practically abuse.

If we ever stumble across what appears to be a fountain of pure parenting wisdom—the one and true way—this is what we'll find instead: a fountain of pure parenting contempt.

18

The Murky Origins of Bipedalism, or The First Toddler

Whether infants are shaken dry by their ankles or swaddled in a hooded duckie towel, the eccentricities of early movement are all sanded down in the end. There's a word for this: *equifinality*. Motor development takes many different paths but it always ends at the same place: we all get up and walk.

In our human-ruled world—in which we can go through a day and see no species but our own, plus a few squirrels—bipedalism is too ubiquitous to notice: I'm bipedal; you're bipedal. Who isn't bipedal, really? We don't stare out the window in the morning and shout, *There are things out there getting around on two legs.*

But maybe we should. To everyone but us, being bipedal is deeply, deeply abnormal. Humans are the only primate species—out of over two hundred—that habitually walks upright. Give or take a few quasi-contenders, we are the only mammals to do so.

Bipedalism is fundamental to what made humans human. In the fossil record, it shows up millions of years before tool use or the explosion in brain size. Long before *Homo sapiens* came along, some of our ancestors were getting around upright, along with a lot of other creatures who didn't turn into our ancestors—according to the archaeological evidence, there was probably a wild array of erect hominids.

The new physiology of bipedalism allowed for the evolution of far bigger brains. It bent the birth canal into a nearly inescapable maze. It ultimately altered the entire body: the feet arched and lost their opposable toes, the head was pulled back over the spine, the pelvis flattened and stabilized. It's hard to find a part of the body that wasn't remade by having to bear the weight of being upright.

A minor example: the gluteal muscles—what you are sitting on—are the engine of chimpanzee movement. But once upright, early hominids no longer needed the gluteals for power: they needed them for balance instead. So the gluteals were enlarged and shifted. "If not for the realigned gluteals along the sides of your hips, holding you steady as you begin to list to starboard or port, you would be thrown off-kilter as you struggled to maintain balance," the anthropologist Craig Stanford writes. You would look, in other words, a lot like a chimpanzee: when chimps stand, they tilt back and forth, compulsively, like a child who really needs to pee. Every walking step is a balancing act: for part of each stride, we are on only one leg. In that one-legged instant, the gluteal muscles are what give us grace. Chimps stumble. Humans stride.

The temptation to speculate about the big question here—*why*—is irresistible, and since Darwin, that temptation has been resisted by nearly no one. Darwin argued in *The Descent of Man* that early hominids stood up so they could use their hands for hunting. In the almost 150 years since, his theory has been joined by a motley assortment of others—the thermoregulatory hypothesis, say, which holds that being upright made early humans less exposed to the intense heat of the savannah (and thus able to spend more time looking for food); or the somewhat more eccentric sexual selection theory, that being upright allowed our male ancestors to more fully display their sexual equipment. The most compelling explanations, though, are piecemeal: they depend on small shifts in circumstance or opportunity.

In certain circumstances, other primates will rise onto two feet—orangutans, gibbons, chimpanzees, bonobos all walk, and even run, when it suits them. In Tanzania, Stanford once tracked a group of chimpanzees through the forest until they reached a stream:

> Chimpanzees don't particularly like water; they're fascinated by it but avoid getting in too deep. This chimpanzee, a big male, waded through. As he reached the other side a few yards away, with water swirling around his legs, he stood up. Looking for all the world like a swimmer emerging from the surf, he reached up, plucked some leaves, stuffed them in his mouth, and then, grasping a plant for support, stepped out of the streambed into the forest.

Chimps should carry a sign: WILL STAND FOR FOOD. Stanford's theory suggests that, millions of years ago, bipedalism may have emerged as a way to get more food, on the ground and in trees. As forests gave way to grasslands, the food supply scattered, giving anyone with a better mode of locomotion, however haphazard and accidental, an advantage. With walking came meat, the nutritional and caloric boost bigger brains needed, fueling further advances.

Any true gains in efficiency were accrued with painstaking slowness, over millions of years. The earliest bipeds, seen from our perspective, must have been comically bad at getting around, even worse than toddlers. We have reaped the rewards: walking upright saves modern humans a massive amount of energy. The evolutionary biologist Daniel Lieberman—who also argues, in yet another bipedal theory, that our physiological peculiarities are adaptations for running, not walking—has conducted experiments on the relative efficiency of humans walking upright versus apes walking on their knuckles. Humans win: bipedalism is four times more efficient. The upright stance has now been honed so sharply that standing up requires only 7 percent more energy than lying down.

More than any other characteristic, walking is what shaped us into human beings. It's our most profound inheritance.

But we can still discard it.

In the herky-jerky world of infant locomotion, there's a type of crawling that has long stood out for its unlikely grace and speed: bear crawling. This sort of scampering—hands and feet on the ground, knees and back off it—sounds ridiculous, but it turns out to be strangely efficient: toddlers who hit upon this style early on often stick with it for months or years, even after they learn to walk. The bear crawl somehow seems to be heritable, too: it runs in families.

The early anthropologist Ales Hrdlicka was so fascinated by the phenomenon that he published a book on it—*Children Who Run on All Fours: And Other Animal-Like Behaviors in the Human Child*—back in 1931, consisting mostly of correspondence with the parents of bear walkers. "I am so glad my six 'monkeys' are of interest to you, for I always insisted that it was interesting for a whole family to run about like that," wrote a mother from Tennessee. A man from Chapel Hill wrote about his nephew, who could walk perfectly well but loved to trot on all fours and "would cover ground at a rate somewhat faster than a man would ordinarily walk; he never seemed to tire." The accounts were not always strictly on topic—"As an interesting incident may be mentioned that once in running on all fours he picked up an apple with his teeth. Sincerely, Chester L and Mrs. Fordney"—but Hrdlicka dutifully tabulated them ("Character of Walk on All Fours," "Additional Peculiarities"), reprinted a few photographs of the bear crawl in the wild, and essayed a few conclusions. The basic cause, he wrote, "is apparently of atavistic nature, the whole phenomenon being thus one of the order of functional reminiscences of an ancestral condition." Which means, basically: they're acting like apes.

The subject of bear crawling hibernated for many decades until a couple of Turkish doctors, in 2004, made a discovery that was more science fiction than science: in a rural village, they happened upon a group of siblings who *had never stood up*. Members of a family with nineteen children, all of whom bear walked in their infancy, these five brothers and sisters had never lifted up off their hands. They had walked like bears all their life.

Unlike the knuckle-walking of chimpanzees, and the bear crawling of babies and toddlers, the siblings wrist-walked, their palms pressed flat against the ground. (Think of someone doing the yoga pose downward-facing dog while walking.) No one had ever seen an adult human move like this before.

The siblings were able to stand upright if they concentrated on it, an early report on them noted, "but they become unsteady if they try to walk bipedally, and soon go down onto their hands." They were quadrupeds. To help support the family, one lone male bear-crawler ranged as far as a mile from home collecting cans and bottles. Bear crawling, he was nearly indefatigable. "This contrasts markedly with normal adult humans," the authors of the report wrote, "who find such a gait—if and when they try it— tiring and uncomfortable even after practice."

The siblings all had a poorly developed cerebellum, the area of the brain that controls for balance, but some humans with no cerebellum still walk. So why did these siblings never stand up? Hrdlicka presumably would have argued that they were reverting to "an atavistic nature," and the Turkish scientists did too: they contended that the siblings were a case of "backward evolution," a missing link to our quadruped past.

There's a less outlandish explanation, though. The bear crawl was efficient enough that if the siblings had floundered at walking early on, and they likely did, they might have just given up on it. They lived in a rural village and kept to their own family. Their parents had accepted their children as they were; they'd never tried to teach them to walk. In this very small world, walking on hands and feet made nearly as much sense as walking upright. They had created their own culture. And in fact the Turkish siblings were always capable of walking: after their story made the news, they received motor therapy. They became bipedal.

A perfectly healthy child will always be bipedal. But there are many children with serious neural impairments, and the story of the Turkish siblings suggests if these children were left alone, at least some might not find their way to their feet. Without social pressure and parental encouragement—*just one more step, honey, one more step*—the Turkish siblings might not be nearly as weird as they seem. Indeed, only a couple of years after the siblings were found, they were no longer complete anomalies: a family with three quadrupedal brothers had been discovered in Iraq, followed by another Turkish family and a couple of families in South America. The bipedal human gait may be our most profound inheritance and yet still be more flexible, less hard-wired than we assume. Despite its constraints, despite its usual predictability, development is a creative, highly sensitive process: it still has, in some isolated corners of the world, the capacity to surprise us.

Toddlerhood

I could simply say that Isaiah first walked when he was fourteen months and we were visiting Anya's brother in New York City, and he squirmed away one afternoon and just *went for it*.

But I'd prefer more of a Great Baby approach to his history:

Isaiah took his first steps in the early evening of a weekend in February 2010, in the living room of a one-bedroom apartment in the Inwood neighborhood of Manhattan, a few blocks down from where the Dutch purchased the island from the local Native American tribe, 374 years before.

This is the end of infancy, after all. It needs some trumpets.

Babies never rest on their laurels: on the way to walking, they constantly trade the known for the unknown, competence for incompetence. It's only when they stand up that their orientation is set: they have achieved their bipedal destiny. It's atypical that they're not asked to give *that* up. It'd be in keeping with their whole lives thus far to have to abandon walking and learn to fly instead.

When Isaiah stood up in Manhattan, a pint-sized King Kong, he faced the world in the posture of the rest of his life: erect, bipedal, unsupported.

He managed three, four, maybe five steps at a time. His eyes were like dying stars: they exploded.

He growled the entire time, a deep, back-of-the-throat *rarrrr* of terrified exhilaration. Every single part of his body was in motion. The only parts not in constant, undulating motion were the very parts that were supposed to be in motion: his lower half. It was as if the sheer effort of keeping his legs steady had displaced a whole body's worth of mad wiggling to his other appendages. His arms were a roller coaster; his mouth was stretched taut, as if it were yet another body part he had to keep under control. He fell not because he lost his balance, although he did that, too, but because it seemed like his legs couldn't stay stiff any longer. He didn't fall; he sort of folded in on himself.

An academic point on stiffness: the human body is really just a collection of things linked to other things, which means it is really just a collection of places for things to go wrong. The Soviet physiologist Nikolai Bernstein pointed out, back in the 1930s, that there are simply too many parts in the body—joints, muscles, bones—for them all to be coordinated individually. There are too many "degrees of freedom," in Bernstein's phrase.

Instead, there are certain patterns of coordination that appear again and again—certain groups of muscles that are moved. Bernstein's insight was that we don't move muscles for the purpose of moving muscles; we move them *to do things*. A person can legibly sign his name on the blackboard using a broomstick, he pointed out. That involves very different muscles than signing your name on a check, but the result is recognizably the same signature.

To reduce these degrees of freedom, new walkers move in a stiff, partially defrosted manner, like baby Frankensteins, as if they are trying to limit the number of body parts in motion at once. Over the next couple of months, Isaiah went from Excited Frankenstein (stiff-jointed, hands up) to Frankenstein (stiff-jointed, hands down) to boy. He began a machine and ended up a human.

Soon his feet hit the ground from heel-to-toe, each arm swinging in time to the opposite foot. Like most new walkers, he got much better in his first six months and then he hit a wall. Kinesiologists will tell you that children don't walk with adult fluency until seven years of age.

In the photos we took of him walking in Inwood, there's only one that caught my face. I'm sitting on a recliner, a few feet away, slumped forward, my hand on my forehead. I look *wiped*. Isaiah vibrates in the foreground: bipedal, arms up, mouth open, his whole body *alive*. It's a bizarre juxtaposition.

I remember that what was going through my head was something articulate like, *Whoa*. It was a lot of exhausted delight, a little bit of sadness, and a sense that things were going to be different now, and not just because he'd be harder to catch.

After language, there is no development as profoundly life-changing as walking. Language, for that matter, develops very slowly, and in different ways (receptive, spoken, intelligible). Compared to it, walking arrives in a rush, bringing with it a whole new way of being in the world.

The psychoanalyst Margaret Mahler, back in the 1970s, argued that walking marks "the psychological birth of the infant." Locomotion allows a baby to separate herself from her parents—and as importantly, it allows her to rush back for shelter, and then wander off again. Infants can now "check in" with their parents, looking back across the room for approval or guidance or just to make sure someone is still there. After months of compliance, movement brings opportunity and the possibility of disobedience. It gives a baby a heady whiff of willfulness.

Some of these emotional changes are sparked earlier, with crawling. Babies, when they begin to move, suddenly register more intense emotions. They are angry more often and more deeply, and they are more determined, more headstrong. They also display more outbursts of sheer joy.

But mothers change, too, and dramatically:

> Mothers of crawlers stated that they began to expect compliance from their infants; they felt that their infants were now responsible for their actions, and hence, they were expected to obey the mother. Mothers of locomotors also reported increasing their use of verbal prohibitions and mentioned how they used their voice predominantly as the means of conveying prohibition. Most strikingly, they reported a sharp increase in their expression of anger toward their infants, stating in many cases that it was the first time in their relationship that they had been angry toward the infant.

Almost unconsciously, mothers begin to demand, in exchange for independence, obedience. As if in recognition of the poignancy of this trade, they also show more *positive* emotion—they hug their babies more often and more intensely. They intuit that a child who walks is a child who sees less of his mother.

The babies seem to sense the very same thing. A study conducted on infants before the onset of walking found that the "motivation to engage" with objects—carrying objects, offering them to their parents, going across the room in pursuit of them— successfully predicted whether a baby would be walking a couple of months later. The more interested an infant was in his surroundings, and in sharing what was in his surroundings, the more likely it was that he'd start walking soon. In other words, walking may be sparked not just by neural developments, and bodily changes, and experience. It may also be sparked by sociability: babies walk because they want to hang out with everyone else. They want to be more fully in the world.

People study infancy because it promises to be about more than infancy. That's why academics pay attention to babies, and that's why parents, at least sometimes, pay attention to academics. We want to know how to raise a happy, smart, confident, curious, kind, thoughtful adult. At a minimum.

But walking is singular. It's not like other developments in infancy. Walking isn't preparation for the wider world. Walking *is* the wider world.

This isn't infancy anymore. It's toddlerhood.

Coda, or A Bedtime Story

No other animal species has been cataloged by responsible scholars in so many wildly discrepant forms, forms that a perceptive extraterrestrial could never see as reflecting the same beast.

> —William Kessen, "The American Child and Other Cultural Inventions"

"How do we know," says a cautious mother, "that today's theory is a better theory? These theories begin to resemble women's fashions. How do we know that clock feeding won't come back next year along with a change in the hem line?"

> —Selma Fraiberg, *The Magic Years*

I started reading about pacifiers because I wanted to understand my own peculiar unease—I had no idea anyone else was uneasy, too.

I didn't have any intention of writing a book that began with sucking. But I was spending so much time watching a baby suck that it began to seem like the only logical subject. Not just logical: obvious. Unavoidable. What else would a book be about?

And if you can start with sucking, you can start almost anywhere. Why not spitting up? Why not spitting, shitting, screaming, sharing? As if pitching that very book, the other day Isaiah stuck his head in the toilet and screamed, *"Poop!* You forgot something, Poop! You forgot the toilet paper!"

The fact that you can start almost anywhere says a few things worth underlining, as if this were the sort of book that provides underlined bits of take-home advice:

1. The most mundane things that a baby does are interesting. The sheer density of development in infancy renders every bare act meaningful.

2. Even if they weren't interesting, adults would have managed to make them interesting. Over time, parents have barnacled the most routine activities in infancy with their own preoccupations. It's sometimes hard to see the baby for all the barnacles.

There's a tension just beneath the surface of this book: it is, very roughly, the friction between numbers 1 and 2—between how babies behave and how adults behave.

In the last couple of decades, scientists have made staggering advances in their understanding of babies. These aren't minor accomplishments. Infancy research isn't like trying to find a needle in a haystack. It's like asking the haystack to tell you where the needle is.

This deluge of new knowledge is extraordinary. But even if the deluge continues—even if we are up to our ears in infancy studies—it will never be enough.

That's because being a parent is a constant search for stories that make sense. We desperately need stories to explain these inscrutable creatures to us, to interpret their mystifying behavior, to make it culturally comprehensible.

There are stories about seemingly every sphere of infancy. Sucking, smiling, touching, toddling: these are extremely mundane activities. They turn out to be unexpectedly intriguing activities—who would have thought that the development of locomotion would turn out to be a *creative* process? But they are nonetheless so routine they can almost be overlooked. And yet, over the ages, we adults have managed to construct a library's worth of stories about activities this banal.

This search for stories has taken parents to very different places. At the beginning of the last century, many parents in this country believed you should touch your child as little as possible; at the end of it, many believed you should touch your child as much as possible. The child hadn't changed much during that time. What had changed were the parents.

More knowledge of infancy, even a flood of it, won't supersede our need to find stories to tell about our children. And those stories will keep changing for a simple reason: their content is as much about the grown-ups as the infants.

Even the professionals have trouble telling the two apart.

In 1960, the French historian Philippe Ariès published the book *Centuries of Childhood*, the cornerstone of a new historical discipline. Childhood in the Western world, Ariès argued, was a recent invention—through the Middle Ages and beyond, children were seen simply as smaller adults. Childhood essentially did not exist. Parents did not really care for their children; the children were just, you know, around. Ariès's vision of childhood was hugely influential: it held sway for a generation.

Then the backlash arrived. New scholars, working deep in the archives, returned with overwhelming evidence that adults could tell the difference between themselves and babies, and that the parents of the past did in fact care deeply about their children. They may have treated them very differently—caring for your child may be a constant, but *how* you care for your child clearly isn't—but the adults of the past weren't oblivious to them. These perspectives are fundamentally different, to say the least, and this scholarly tug-of-war hints at just how hard it is to excavate the history of childhood. Children leave few records of their own: a coral ring here, a walking stool there. Not only do the adults write the histories, the children can't even read them.

The babies are murmurs of the future, drowned out by the noise of the present.

All this is why, as the historian Carolyn Steedman has observed, "the claims for a history of adult attitudes toward children, and of upsurges of interest in children and the notions of childhood they embody, are much more compelling than claims for a history of children." In other words, all histories of childhood are ultimately histories of *adulthood*.

That includes this book, of course. You start out writing about babies and you end up writing about yourself.

It seems inevitable that historians would suffer from the my-baby-and-me problem. Historians, like parents, are in the storytelling business.

Scientists, on the other hand, deal with data. It is harder, colder, less malleable. But even data often ends up looking a lot like whoever is looking at it.

There are a few camps in developmental psychology, and as in a thinly veiled roman à clef, each camp's account of the infant sounds like each camp: psychologists who rely on theoretical models argue that the infant is fundamentally a theorizing creature. Psychologists who rely more on observation argue that the infant pieces together the world through observation. The infant, *c'est moi*.

A couple of years ago, the writers Po Bronson and Ashley Merryman published a book about a wave of counterintuitive child development research—praise ultimately discourages children, not talking about race makes racism more likely, that sort of thing. They called it, ominously, *Nurtureshock*, and it was packaged as the future of child-rearing books: no longer did parents have to subsist on vague, subjective advice; they could now have expert-honed, blind-trialed truth.

Reviewing the book in *The New York Times*, the writer Pamela Paul saw it as just another swing of the pendulum of parenting advice. Each generation kills the parenting gurus of its parents. *Nurtureshock* was simply a return to the century-old child study movement and the belief that child rearing could be perfected through rigorous, scientific analysis. It wasn't Truth. It was Fad.

Nonsense, Bronson and Merryman responded. Their argument was, basically: That old science was lousy! This new science is better!

This is actually a decent point: the experiments *Nurtureshock* featured are more sound than the old child study–era experiments. These days, almost all experiments are; things are just more rigorous than they used to be. John Watson's science really was worse science.

And yet. And yet. There are a few things to add here. First, science is born out of culture: John Watson was trapped in his and we're trapped in ours. Our cultural beliefs shape how experiments are conceived, carried out, interpreted, translated for the popular audience. This problem is aggravated in infancy research: the infant is mute; we have to voice it. We now do so in very sophisticated, statistically significant ways, but we still voice it.

The second is that these experiments have the most meaning

within the discipline, not outside of it: they are mostly relevant to small academic disputes, not large parenting decisions. But when we extract practical advice from these studies, we shear off all their disclaimers and complexities. These are often experiments that show real but very small effects, but in the child-rearing advice genre, a study that showed something is *possible* comes out showing that something is *certain*. Meager data, maximum conclusions.

The last is that the science is, almost always, in no way settled. Developmental psychologists can't tell you how to raise your children because developmental psychologists wouldn't be able to agree on how to raise their own children. It's a discipline with crevasses under every certainty. The biggest crack is between psychologists who are primarily interested in *origins*—what abilities infants may or may not be born with—and those who are primarily interested in *process*—the question of how these abilities develop. As the study of infancy has become more sophisticated, the apparent abilities of infants have become more sophisticated, too. It can seem as if every new experiment claiming babies are capable of a task at six months is followed by another claiming that they're capable of it at *four* months. To put it crudely: psychologists interested in origins, or innate knowledge, consider these early capabilities crucial; psychologists interested in development consider them a distraction. The camps even disagree on whether the key experimental paradigm in infancy research, the looking time or habituation paradigm—the way researchers tell if babies have noticed something new—actually measures what it claims to measure.

In other words: even if I'd still written about the same activities—suck, smile, touch, toddle—*this whole book could have been different*. Take the stunning research Andrew Meltzoff did on neonatal imitation. Meltzoff's work has been replicated many times, and for him and many others, the issue of neonatal imita-

tion is settled. But is it? It's widely acknowledged that tongue pro-
trusion is the gesture infants most reliably imitate. The psychologist
Susan Jones, whom we last saw in these pages in the "bowling
alley" infant smiling experiment, has conducted experiments
suggesting that this "imitation" is coincidence. Jones found that
newborns stick their tongues out in response to almost *any* interesting visual stimuli. Why? Because their tongue is their world:
they have almost nothing else.

Meltzoff's account of neonatal imitation is still the conventional account. But it is hardly undisputed. When someone cites
science as a guide to child rearing, they're citing someone's *version*
of the science. Any book about science, including this book, takes
someone's version sometimes and someone else's version other
times. Books about child development are usually at war with
themselves: the experts inside can't stand to be between the same
covers. This is disconcerting. The last century of Western parenting is the story of parents exchanging folk knowledge for expert
knowledge. When the experts start yelling at each other, it begins
to look like a poor trade.

We want to believe we're getting better as parents. Think of
the steady accumulation of expert scientific findings—we *must* be
improving. We want to believe it so much, in fact, that we should
be deeply suspicious when anyone tells us there are new and immediately useful insights to be found in developmental research.

When people say things like this, in short:

Backed by more than 1,200 trusted sources, *The Baby Bond: The
New Science Behind What's Really Important When Caring for Your
Baby*, by parenting expert and . . .

More than 1,200 trusted sources! But why not 1,300 sources?
Wouldn't 1,300 be better? I mean, how can you really trust only
1,200?

What's a Parent Got to Do with It

Seemingly every culture before our own has had a single acceptable way to raise a baby. These cultures wouldn't have cared about
the new scientific findings: they already knew how babies worked.
Their answers were all very different, mind you, but they had this
in common: all the other answers were wrong.

Such confidence makes sense. If you have to raise a baby, not
study a baby, you'd better settle on an answer, and as long as you
have settled on an answer, you may as well be certain about it.
Pretty much everyone has been very certain. But if everyone has
been very certain, and everyone's certainty has been very different, you start to suspect that there aren't that many certainties
after all. There's no one true path. Or put another way: the one
true path is forked.

A lot of what we assume to be essential to infants are in fact
cultural fictions—just-so stories to get the parents to sleep at
night. That can be hard to believe in the abstract. So let's take a
fundamental characteristic of Western parenting, something impossible to conceive of not doing: looking your kid in the eye.

Face-to-face interaction is the foundation of modern Western
parenting. It is how we explain what it means to be human—this
is how we show emotion, this is how we communicate, this is
how we make funny flatulence sounds. It's such an instinctive response to a baby that it is almost impossible to imagine parenting
without it: mugging is what babies are made for.

It goes without saying that a parent who didn't interact in this
way would be a poor parent and that his child would be developmentally crippled in some capacity. Remember the German and
the Nso mothers, who were each bewildered by the parenting
style of the other. The lack of face-to-face exchanges among the
Nso dumbfounded the German mothers. The absence was so
strange that the German mothers were "somehow reluctant" to
accept it. It just didn't make sense.

There are good reasons for assuming that face-to-face parenting is equivalent to good parenting. Face-to-face interaction allows infants to see their effect on the world—what psychologists call contingency perception. They begin to understand the idea of dialogue, of consequences, of the self. This very book, in its discussion of the work of Colwyn Trevarthen, has underlined the importance of face-to-face interaction.

But in much of the world, parent-infant interaction isn't considered important at all. Babies are carried facing outward: they don't see their mother; they see everyone else.

To the ethnographies we go: in the Marquesas Islands of the South Pacific, babies face outward; they are supposed to engage with other people, not the mother. And if they don't, they are constantly prompted to do so. Their mothers "discourage babies from becoming self-absorbed, directing them to attend to others. They call their names, jostle them, and tell them to look at so-and-so." From a very early age, these babies are engaged in the social life of the community; their world is broader than the dyadic world most Western infants inhabit. Many infants in the world are raised this way. It's just that very few of those infants live in the West. A study comparing native French mothers with West African mothers living in France found that less than 10 percent of what the French mothers said to their babies was a reference to someone else—someone who wasn't the mother or the baby. But some 40 percent of what the West African mothers said referred to someone else.

Face-to-face engagement, that basic tenet of decent parenting, turns out to be nothing more than a parenting *style*. The French mothers were preparing their infants for a culture in which social life is conducted one-on-one. The West African mothers were preparing theirs for a society of communal engagement. Their child rearing stresses the significance of the community, not the mother. This dynamic holds up: the more important the idea of individuality to a society, the more likely it is to have a dyadic parenting style. Middle-class European mothers, for example, spend twice as much time face-to-face with their babies as middle-class Japanese mothers do.

You can see the imprint of these different styles as early as toddlerhood: toddlers raised with lots of face-to-face time recognize themselves in the mirror at an earlier age—they have somehow grasped the idea of their selfhood sooner. Toddlers raised with lots of physical contact and less face-to-face time are compliant at an earlier age—they have somehow learned to regulate their own desires sooner.

It's not just that there's no right way to raise an infant. There may not be any right way to *relate* to an infant, either. We're starting to run out of potential types of right ways.

It shouldn't be surprising that all these *right ways* aren't. The job of culture is to be arbitrary, or at least arbitrary in a culturally sensible and coherent sort of way. Its function is to lay down the law on exactly how you should behave: *this* is how you raise a baby. Historically, these sorts of arbitrary, culturally embedded rules for child rearing weren't a shortcoming. They were the whole point. They kept you from going crazy.

Today we live in an era in which you can raise a baby a lot of different ways. You can choose your own adventure. It's a uniquely modern problem. It might not seem like a problem at all—it might seem like the opposite of a problem—but it has cul-de-saced in a paradox of plenty. How do you decide what adventure to choose? There are too many yogurt brands in the yogurt case.

We have more information and more choices and they have left us more confused. We can't go back to less information and fewer choices. Yet we still desperately want to know the *right way* to raise our children.

The problem today is that no single system looks right. But that might be the solution, too: if everything looks a little wrong, we might get past the need to be wholly right. Which is why the single most sensible child-rearing manual I have ever read, *A World of Babies,* is a book of child-rearing advice that is an implicit attack on the idea of rightness. Each chapter is a miniature child-rearing manual from a different society—the Ifaluk of Micronesia, the Fulani of West Africa—in the voice of a Dr. Spock–like figure from that society. (The actual authors are academics who study each society.)

It's a brilliant conceit. The book's tone is instantly familiar—it casts the hypnotic, reassuringly authoritative spell of any conventional child-care guide—but the content is utterly foreign. It feels as if you're peacefully reading T. Berry Brazelton and then he tells you to coat your child in cow dung to fool the witches. You can never read Brazelton the same way again.

The Importance of Being Worried

Somehow, and despite all the cow dung, the babies do all right.

The psychologist Jerome Kagan has argued that parenting has a threshold function: up until that threshold is crossed, the effects of a child's very early experience even out in the end. But parenting that crosses the threshold—abuse, stress, utter indifference—can sink in deep, especially if the baby remains in that environment. There's a lot to be said for this perspective on parenthood, not least that it offers well-meaning parents some relief from scaremongering.

It also accounts for the astounding flexibility of the human infant: he is game for the craziest parenting stuff you can come up with. There may be no more extreme contrast in parenting philosophies, as Melvin Konner once observed, than that between the Kung and a traditional Israeli kibbutz. The Kung, of course, are in extremely close contact with their infants: they are nursed constantly, carried constantly, slept alongside. In a traditional Israeli kibbutz, children are raised communally: in early infancy, babies sleep in an "infant house" with other babies. They spend extended periods of time with their parents, and are close to them, but their nights and much of their days are spent apart.

It's an extraordinary juxtaposition: it is hard to imagine parenting strategies more different. And yet, as Konner observed, "children in both settings grow up to be basically normal psychologically." Since Konner made this comparison, a few decades ago, the kibbutz model of communal child rearing has virtually disappeared. It failed for a variety of reasons, including the most indelible—the parents missed their children. But it didn't fail because the children turned out badly. They did just fine.

This is a very hard lesson to learn.

In the lineage of child-rearing philosophies, we modern parents would seem to be Rousseau's descendants. Rousseau, and then Wordsworth—who wrote, immortally, that babies came straight from heaven to us, *trailing clouds of glory*—fleshed out the child with soul and spark and naïve genius. Somersault your way through it, they said of childhood: it will soon pass. For all his praise of the innocent joys of childhood, though, Rousseau also wanted to manage it, to control what children saw and encountered, so skillfully and subtly that children would be simultaneously free and controlled. This was his self-contradictory bequest: managed play. In many ways, Rousseau's vision of childhood is so much like our own that it can seem inevitable that we inherited it.

But it wasn't. A half century after *Emile,* Rousseau's vision— that children should be raised wild and free, uncorrupted by worldly society—looked like an idle, impractical daydream. A new age of industrial capitalism was dawning, and it was awkward to conceive of children as exceptional and innocent and also fit for factory labor. But just as important, Rousseau's romantic idea of childhood simply didn't make sense to most people. The popular English writer Hannah More wrote that it was "a fundamental error to consider children as innocent beings, whose little weaknesses may, perhaps, want some correction, rather than as beings who bring into the world a corrupt nature and evil dispositions."

This sounds brutal today, but it wasn't an atypical attitude, post-Reformation. In Catholicism, baptism washed away original sin: a baptized baby was an innocent baby. In new Protestantism, baptized babies remained sinful: they could be, and likely were, evil. This is America's original child-rearing philosophy; it arrived on the *Mayflower*.

This belief is also why many strict Protestants, and especially the Puritans, were so attentive toward children. "The Puritans were obsessed with children," the historian Steven Mintz has noted. "They regarded children as a trust from God and the key to creating a Godly society. In secular form, this deep concern remains the American attitude today." The Puritans thought more deeply about child rearing, and wrote more about it, than anyone had before.

We think like Rousseau. But we worry like Cotton Mather. "If we are looking for a time when the level of parental anxiety about children matches that in the early twenty-first century," the historian Hugh Cunningham writes in *The Invention of Childhood*, "it is perhaps among the Puritans of the sixteenth and seventeenth century that we will find it." The Puritans, for all their indelible otherness, are ultimately our child-rearing forefathers. We obviously worry about very different things. But our sheer quantity of worry unites us. It's possible that no people before or after the Puritans ever invested more in their children. Until us—we Puritan moderns.

"We have encountered anxious parents in the past," Cunningham notes. "We have seen them tormented by their children's illness or death, worried about the best way to bring them up, anxious about their behaviour when they leave home. But there has probably been no previous generation of parents that has been quite so constantly concerned for their children and their future as our own."

For as long as they have been having them, it seems safe to assume that human beings have worried about their babies. This was a rational reaction: for thousands of years, infancy was a gauntlet of deadly dangers. It was a stage of life that parents were relieved to see end. No one then stopped anyone in the grocery store and said, *Enjoy every precious moment: it all goes by so fast.* That's not to say parents weren't delighted by their babies: they often were. But that delight was darkened by foreboding.

We look down at this fateful history from the heights of modern medicine, knowing that we modern parents will never have to walk through that perilous valley. We expect our children to live out infancy. We are at the tail end of a decline in infant mortality that began just over a century ago. Babies no longer wander into open hearths or are mauled by marauding pigs. We have vaccines, lead-free educational toys, diapers that can sop up a typhoon. But we have never been more worried.

The Puritans had good reasons to worry. But the future of a newborn today has never been more assured. We occupy a privileged place in the history of infancy: we are free to enjoy it.

No one ever said enjoying it came easy.

Back before Isaiah could sit up, he very occasionally had a Bumbo do his sitting for him. The Bumbo is a spongy, brightly colored seat that holds an infant upright. It's basically the same thing as those seats some traditional societies make out of dirt or sand, except the Bumbo's made out of molded foam (modern baby accoutrements being mostly a way of charging for dirt and sand). One afternoon, Isaiah was sitting, if that's the right word, in the Bumbo in our backyard, when our genuinely lovely German next-door neighbor, the mother of a couple of girls, walked past the fence.

"Oh," she said brightly, pointing to Isaiah in the Bumbo, "you know that will crush his internal organs, right?" *This is an exact quote.* We said, Well, *no,* actually, we didn't know that.

She was wrong, obviously. Isaiah's internal organs are in exquisite shape. And as far as I can tell, the Bumbo doesn't crush anyone's internal organs. In fact, a recent Consumer Product Safety Commission alert about the Bumbo says absolutely nothing about internal organs. It only says that sometimes babies tip over in the Bumbo, and sometimes when those babies tip over, they fracture their skulls.

But that's it! Nothing about internal organs! Just craniums!

If parenting today is like a choose-your-own-adventure book, this is a choose-your-own-moral story. Is the moral here that you should worry less or more? On the one hand, our neighbor was warning us about an imaginary danger. On the other hand, while she was warning us about an imaginary danger, we were overlooking a real danger. On the one hand, how many fractured skulls are we actually talking about here? On the other hand, fractured skulls!

Eventually, to stay sane, a parent in modern America has to stop and slowly repeat the wisdom of the great shtetl philosopher Tevye: *there is no other hand.*

> Isaiah: Everything is all messed up!
>
> Anya: Oh yeah? What do you mean? What's all messed up?
>
> I: Everything is all messed up!
>
> A: Like what? Give me an example. What is messed up?
>
> I: Like my clarinet!
>
> A: Your clarinet?
>
> I: My clarinet.
>
> A: But sweetie, you don't *have* a clarinet.
>
> I: *Why?*
>
> —Isaiah, 2 years, 11 months

You can see infancy as a set of skills to master so you can get to the next set of skills. This is the natural way to see it, really: no one wants to be stuck at infancy. Plus, newness here can stand in for value, which solves the problem of how you measure value: more newness is de facto better development. Faster equals better. The best way to become a grown-up is to act like a grown-up as soon as possible.

Among the problems with this perspective is that it shortchanges the actual experience of infancy. Being an infant isn't important; what's important is getting *past* infancy.

Before Isaiah was born, we were already impatient for him to get past infancy. At some point when Anya was pregnant, I remember us talking about our fear that we'd find our baby, well, *boring*. We assumed we would be besotted. We were afraid we'd be besotted and bored. We were afraid of the long barren stretch of days and nights before language. Imagine having a roommate who gave you the silent treatment for a year, we thought.

As it happened, language for Isaiah came late. This wasn't a complete surprise. In Anya's family, language has a history of being tardy. (Her paternal uncle didn't say a word until he screamed that the cow nearby was going to gore him. He was three.) Isaiah moved toward language at twelve months. He said a couple of words. Then he retreated. He didn't speak a word for the next half year. We would have counted almost anything as a word. There was nothing to count.

Humans are creatures of language: we're naturally biased toward it. But there's a lot that happens before words. That was the single most surprising thing about Isaiah: how much of *him* there was before language. The literature of infancy does not lead you to expect so much *himness*. His personality was there before words. When the words came—and when they came, they poured—he stayed the same boy.

When we talk about how important infancy is, we miss the point. We should talk about how *interesting* it is—how a newborn can become himself before he can say a single word. It's an enthralling period of life for its own sake, not just for what it means for adolescence and adulthood. If we always see infancy as a narrative of progress, we never see the baby before us; we always see the child to come.

If we find infancy interesting—if we manage to be more curious than worried—we end up providing the few things that infants actually need: attention, affection, security. The things you can't strip away from the squirrely word "love."

Of course, we don't have babies because we want to have babies; we have babies because we want to have children. And sometimes what's interesting in infancy is most interesting in light of the future. Take a baby's smile. It's glorious in the present, but it is most *meaningful* because of what comes next—because a smile is how a baby begins to participate in the social world.

Having a baby means having to pay a lot more attention to the future. But among the many quiet, scattered points of this book is this, which I am prepared to be loud about: there's a lot to pay attention to in the present, too. Take toddling this time: babies don't learn to walk by walking; they learn to walk by failing to walk, which is happening pretty much all the time. Every minuscule activity of an infant is potentially interesting. This is an attitude that would only make sense in infancy. No one, outside of a very patient therapist, would ever care about every minuscule activity of an adult: your life is boring. Any infant has a more interesting life than you do.

Once again: you start out writing about babies and you end up writing about yourself. This is as close as I get to advice in this book, I think, and inevitably, it is advice I'm giving myself. That's because sometime between when I finish writing this and when you read it, we will have added another small primate to the household.

I like to think I know how I have changed as a parent since Isaiah was born, partly because of raising him, partly because of the experience of this book. I like to think that I'm somewhat less jittery. I'm more confident in the baby, if not in myself. At the very least, I now know that for a long time parents have been doing things to babies that seem like *really bad ideas* and everyone pretty much turns out okay.

These days, okayness isn't much to brag about. Okayness is a fate to dread. The whole point of reading books about child rearing is that they're supposed to help you transcend okayness. In infancy, there's a lot that can go wrong, and very occasionally does: some infants really are raised with neither attention, nor affection, nor security. But it isn't at all clear if, besides those things, there's any more that can go *right*. You can't tiger-mom a one-year-old. Okayness is the best we can hope for.

In the end, I have no idea how much I've changed. To think otherwise would be hubris. After all, I don't have a baby anymore; I have a child. When he wants something, he says it. What he says is not always in the realm of sense—hence: *but sweetie, you don't have a clarinet*—but he makes his own sort of sense. The speechless infant requires you to bring the sense.

Like most people of my generation, I had no real experience of babies before I had my own. I had no sense to bring. I didn't know how to hold a baby. I didn't know how to change a diaper. I didn't know how to burp or comfort a baby. I'd never rocked anyone to sleep; I'd never sung anyone lullabies. In her book *What I Don't Know About Animals,* Jenny Diski calls herself post-domestic—born after the era when people had routine contact with animals who weren't pets. I felt not just post-domestic. I felt post-infant.

When you have no real experience of babies, babies themselves are no help. For a long time after birth—a distressingly long time—a newborn is an inscrutable creature. More than inscrutable, really—*other*. A glorious, unfathomable mystery. The philosopher Thomas Nagel famously asked the question, What is it like to be a bat? His answer was that even if we know everything knowable about bats and about how bats work, that knowledge is limited by our very human minds. It amounts to knowing what it is like to be a bat *if you're a human*. It doesn't have anything to do with what it is like if you're a bat.

Nagel's question was a philosophical question, of course. He didn't actually care what it was like to be a bat. But we care an enormous amount about what it is like to be a baby. And even though newborns are far more knowable than any bat, they still emerge not as a bat but something of a sphinx. There is still a subjective gap that separates us from them.

This book is an account of the many hops that people have tried to make over that gap. And my response to the next glorious, unfathomable mystery in my arms, I suspect, will be to keep hopping.

Notes

This book is a cautionary tale: one damn thing led to another.

Mostly I was led around the library (or online journals). Occasionally, I picked up the phone: any quotations from Karen Adolph, Daniel Messinger, Colwyn Trevarthen, or Tiffany Field that are not attributed in the notes are from personal interviews.

Where I quoted someone or something directly, I have listed the page number below; otherwise, to save space, I have not. Where I discussed a book and its author in the text but did not quote it, I have not listed it below.

INTRODUCTION

6 "Tickling is bad for children": Angelo Patri, *Child Training* (New York: D. Appleton, 1922), 9.

PART ONE: SUCK

"But so soon as": J. P. McKee and M. P. Honik, "The Sucking Behavior of Mammals: An Illustration of the Nature-Nurture Question," in *Psychology in the Making*, ed. L. Postman (New York: Knopf, 1962), 586.

1. SUCKING: A LOVE STORY

12 A treatise on childhood: J. Gillis, "Bad Habits and Pernicious Results: Thumb Sucking and the Discipline of Late-Nineteenth Century Paediatrics," *Medical History* 40 (1996): 64.

13 Mothers who ignored them: Ibid., 66.

13 "The immediate pleasure value": Margaret Ribble, *The Rights of Infants* (New York: Columbia University Press, 1965), 23.

13 "It may be presumed": Charles Darwin, "A Biographical Sketch of an Infant," *Mind* 7 (July 1877): 288.

14 It was, essentially: Wenda Trevathan, *Ancient Bodies, Modern Minds* (New York: Oxford University Press, 2010), 97.

14 "The professions of obstetrics": Ibid., 98.

15 Linnaeus, as the historian: L. Schiebinger, "Why Mammals Are Called Mammals: Gender Politics in Eighteenth-Century Natural History," *The American Historical Review* 98, no. 2 (1993).

16 Since taste buds emerge: R. L. Doty and M. Shah, "Taste and Smell," in *Encyclopedia of Infant and Early Childhood Development*, eds. M. Haith and J. Benson (San Diego: Academic Press, 2008).

17 But it may be more fair: E. M. Blass and V. Ciaramitaro, "A New Look at Some Old Mechanisms in Human Newborns: Taste and Tactile Determinants of State, Affect, and Action," *Monographs of the Society for Research in Child Development* 59, no. 1 (1994).

17 At a week, breast-fed: Doty and Shah, "Taste and Smell," in Haith and Benson, *Encyclopedia of Infant and Early Childhood Development*.

18 "It is clear from recent lay": M. Woolridge, "The 'Anatomy' of Infant Sucking," *Midwifery* 2, no. 4 (1986): 164.

18 What babies do: C. W. Genna, ed., *Supporting Sucking Skills in Breastfeeding Infants* (Sudbury, MA: Jones & Bartlett, 2008). Also see: A. L. Delaney and J. Arvedson, "Development of Swallowing and Feeding: Prenatal Through First Year of Life," *Developmental Disabilities Research Reviews* 14 (2008); and K. Mizuno and A. Ueda, "Changes in Sucking Performance from Nonnutritive Sucking to Nutritive Sucking during Breast- and Bottle-Feeding," *Pediatric Research* 59, no. 5 (2006).

2. LACTATION AND ITS DISCONTENTS

21 In the most remarkable: A. M. Zivkovic et al., "Human Milk Glycobiome and Its Impact on the Infant Gastrointestinal Microbiota," *PNAS* 108, supp. 1 (2011).

22 It is as if: N. Wade, "Breast Milk Sugars Give Infants a Protective Coat," *New York Times*, August 2, 2010.

22 A recent report: R. Black et al., "Maternal and Child Undernutrition: Global and Regional Exposures and Health Consequences," *The Lancet* 371, no. 9608 (2008).

24 In 1925: J. L. Mursell, "The Sucking Reaction as a Determiner of Food and Drug Habits," *Psychological Review* 32 (1925): 409, 415.

25 The psychiatrist Daniel: Daniel Stern, *Diary of a Baby: What Your Child Sees, Feels, and Experiences* (New York: Basic Books, 1998), 40.

25 "The only emotional": Rebecca Kukla, *Mass Hysteria: Medicine, Culture, and Mothers' Bodies* (Lanham, MD: Rowman & Littlefield, 2005), 208.

26 Here's what happened when: See Meredith Small, *Our Babies, Ourselves* (New York: Anchor Books, 1998); D. B. Jelliffe and E. F. Patrice Jelliffe, *Human Milk in the Modern World* (New York: Oxford University Press, 1978); and Wenda Trevathan, *Human Birth: An Evolutionary Perspective* (New Brunswick, NJ: Transaction Publishers, 2011).

26 For us, breast-feeding: Jelliffe and Jelliffe, *Human Milk in the Modern World*, 2.

27 Long ago, the study of lactation: For more on physicians and percentage feeding, see Rima Apple's *Mothers and Medicine: A Social History of Infant Feeding, 1890–1950* (Madison: University of Wisconsin Press, 1987).

29 So it shouldn't come: See D. Geddes, "Inside the Lactating Breast: The Latest Anatomy Research," *Journal of Midwifery & Women's Health* 52, no. 6 (2007).

29 "This is not to deny": C. von Hofsten, "Motor and Physical Development: Manual," in *Encyclopedia of Infant and Early Childhood Development* (2008).

31 When psychologists tracked how: P. Rochat, "Differential Rooting Response by Neonates: Evidence for an Early Sense of Self," *Early Development and Parenting* 6, no. 34 (1997).

31 But there is now: C. von Hofsten, "Motor and Physical Development," 375.

32 Prolactin is what: Sarah Blaffer Hrdy, *Mother Nature: A History of Mothers, Infants, and Natural Selection* (New York : Pantheon Books, 1999), 127.

33 Colostrum suffered: See J. M. Morse, C. Jehle, and D. Gamble, "Initiating Breastfeeding: A World Survey of the Timing of Postpartum Breastfeeding," *International Journal of Nursing Studies* 27, no. 3 (1990).

34 After Rome, European: Valerie Fildes, *Breasts, Bottles, and Babies: A History of Infant Feeding* (Edinburgh: Edinburgh University Press, 1986).

3. THE VERY WEIRD, PERFECTLY NORMAL STORY OF WET NURSING (PLUS, SOME GOAT MILK AND BEER)

38 The Puritans in particular: M. Salmon, "The Cultural Significance of Breast-feeding and Infant Care in Early Modern England and America," *Journal of Social History* 28, no. 2 (1994): 252–53.

38 The great Puritan minister: Janet Golden, *A Social History of Wet Nursing in America* (New York: Cambridge University Press, 1996), 11.

38 In the eighteenth century, human: Ibid., 27.

Meanwhile, Paris: See George Sussman, *Selling Mothers' Milk: The Wet-Nursing Business in France, 1715–1914* (Urbana: University of Illinois Press, 1982).

39 "What of motherly": E. Shorter, "Review of *Selling Mothers' Milk: The Wet-Nursing Business in France 1715–1914*," *The American Historical Review* 88, no. 3

(1983): 689. For an examination of that very question, see Melvin Konner's essay "Thinking about Maternal Sentiment" in *The Evolution of Childhood* (Cambridge, MA: Harvard University Press, 2010).

40 "Working mothers sent their": Hrdy, *Mother Nature*, 368.

41 "By the sacrifice": Golden, *A Social History of Wet Nursing in America*, 97.

41 Less matter-of-fact: Ibid.

42 In *Emile*, his treatise: Jean-Jacques Rousseau, *Emile* (New York: Basic Books, 1979), 46.

42 The new citizens: Kukla, *Mass Hysteria*, 51.

43 Ironically, as maternal: See P. J. Atkins, "Mother's Milk and Infant Death in Britain, circa 1900–1940," *Anthropology of Food* 2 (2003); and Jacqueline Wolf, *Don't Kill Your Baby: Public Health and the Decline of Breastfeeding in the Nineteenth and Twentieth Centuries* (Columbus: Ohio State University Press, 2001).

43 "A wet nurse is one-quarter": J. H. Wolf, " 'Mercenary Hirelings' or 'A Great Blessing'?" *Journal of Social History* 33, no. 1 (1999): 115.

44 In 1934, an article: K. A. Foss, "Perpetuating 'Scientific Motherhood': Infant Feeding Discourse in *Parents* Magazine, 1930–2007," *Women Health* 50, no. 3 (2010): 303.

45 Like wet nursing, dry: See Mary Spaulding and Penny Welch, *Nurturing Yesterday's Child: A Portrayal of the Drake Collection of Paediatric History* (Toronto: Natural Heritage, 1994).

47 This repugnance was: Fildes, *Breasts, Bottles, and Babies*, 264.

48 It might have started: U. B. Lithell, "Breast-Feeding Habits and Their Relation to Infant Mortality and Marital Fertility," *Journal of Family History* 6, no. 2 (June 1981): 193.

48 Cultures that relied: See Lithell, "Breast-Feeding Habits"; J. Knodel, "Infant Mortality and Fertility in Three Bavarian Villages: An Analysis of Family Histories," *Population Studies* 22, no. 3 (1968); and J. Knodel, "Breast-Feeding and Population Growth," *Science*, 198, no. 4322 (1977).

49 From his home: Fildes, *Breasts, Bottles, and Babies*, 271.

49 The reasons were physiological: P. Cazeaux et al., *A Theoretical and Practical Treatise on Midwifery* (Philadelphia: Lindsay and Blakiston, 1877), 1102.

50 In the 1860 book: Charles Henry Felix Routh, *Infant Feeding and Its Influence on Life* (London: John Churchill, 1860), 328–29.

50 At a hospital in Aix en Provence: Spaulding and Welch, *Nurturing Yesterday's Child*, 70.

50 Infants nursed by animals: Fildes, *Breasts, Bottles, and Babies*, 271.

51 Historical infant mortality: R. Meckel, "Infant Mortality," in *Encyclopedia of Children and Childhood in History and Society*, www.faqs.org/childhood /In-Ke/Infant-Mortality.html (accessed 3 October 2012).

51 Our physiology seems: H. L. McClellan, S. J. Miller, and P. E. Hartmann, "Evolution of Lactation: Nutrition v. Protection with Special Reference to Five Mammalian Species," *Nutrition Research Reviews* 21 (2008).

52 Babies had to survive: Béatrice Fontanel, *Babies: History, Art, Folklore* (New York: Abrams, 1997), 121.

4. TASTY, TASTY THUMB

53 It's been estimated that: P. C. Friman et al., "Characteristics of Oral-Digital Habits," in *Tic Disorders, Trichotillomania, and Other Repetitive Behavior Disorders*, eds. D. Woods and R. Miltenberger (Boston: Springer, 2001).

53 The majority of newborns: Scott Miller, *Developmental Research Methods* (Thousand Oaks, CA: Sage Pub., 2007).

56 In a sort of Kama Sutra: S. Kern, "Freud and the Discovery of Child Sexuality," *History of Childhood Quarterly* 1 (1973): 125.

57 "You might want to counter": Gillis, "Bad Habits and Pernicious Results," 59.

57 Twenty years after: Ibid., 55.

57 "I watched a young mother": "Thumb-Sucking Babies," *Good Housekeeping* (April 1908).

58 "It may be that there are": Samuel Augustus Hopkins, *The Care of the Teeth* (New York: D. Appleton & Co., 1903), 82.

58 For Hopkins, "So hideous": Ibid., 74.

59 "Pediatric textbooks came": Gillis, "Bad Habits and Pernicious Results," 56

59 "If a babe sucks": C. B. Stendler, "Sixty Years of Child Training Practices: Revolution in the Nursery," *The Journal of Pediatrics* 36, no. 1 (1950): 127.

60 "The child," as the historian: Steven Mintz, *Huck's Raft: A History of American Childhood* (Cambridge, MA: Harvard University Press, 2004), 191.

60 A 1923 issue: *Popular Science*, October 1923, 51.

61 "Remember that a baby": C. Gale and C. Martyn, "Dummies and the Health of Hertfordshire Infants, 1911–1930," *Social History of Medicine* 8, no. 2 (1995): 231.

61 A popular English child-care book: Ibid., 250.

61 "Children," Watson insisted: John B. Watson, *Psychological Care of Infant and Child* (New York: W. W. Norton, 1928), 7.

62 "The world would be considerably": Ibid., 12.

62 "Give me a dozen": John B. Watson, *Behaviorism* (1924; reprint, New Brunswick, NJ: Transaction Publishers, 2009), 82.

62 If it continues past: McKee and Honik, "The Sucking Behavior of Mammals," 594.

65 "No one who has": Sigmund Freud, *Three Essays on the Theory of Sexuality* (1905; reprint, New York: Basic Books, 2000), 48. For more background, see Kern, "Freud and the Discovery of Child Sexuality."

65 "Obtaining pleasure is": C. G. Jung, *The Theory of Psychoanalysis* (New York: The Journal of Nervous and Mental Disease Publishing Co., 1915), 23.

66 The psychoanalyst Karl: Karl Abraham, "The Influence of Oral Erotism on Character-Formation," *Selected Papers on Psychoanalysis* (London: Hogarth Press, 1949), 400–401.

66 A special article on: D. S. Palermo, "Thumbsucking: A Learned Response," *Pediatrics* 17, no. 3 (1956).

67 The gentler tone: Martha Wolfenstein, "Trends in Infant Care," *American Journal of Orthopsychiatry* 23, no. 1 (1953): 122.

67 William Preyer: William Preyer, *The Mind of the Child*, vol. 1 (New York: D. Appleton & Co., 1898), 111.

67 The habit: McKee and Honik, "The Sucking Behavior of Mammals," 585–86.

68 "Sucking, and particularly": Ibid., 588.

68 To wit, the article: Joseph Gillman, "Toe-Sucking in Baboons: A Consideration of Some of the Factors Responsible for This Habit," *Journal of Mammalogy* 22, no. 4 (1941): 395.

68 It was not all baboons: A legendary example is David M. Levy's "Experiments on the Sucking Reflex and Social Behavior of Dogs," *The American Journal of Orthopsychiatry* 4, no. 2 (1934).

69 It was an *indication*: McKee and Honik, "The Sucking Behavior of Mammals," 652.

70 "The importance of this": T. Berry Brazelton, "Sucking in Infancy," *Pediatrics* 17, no. 3 (1956): 400.

70 The results, the author: M. Heinstein, "Influence of Breast Feeding on Children's Behavior," in *The Course of Human Development*, ed. M. C. Jones et al. (New York: Wiley, 1971), 303.

71 The simple act of: Friman et al., "Characteristics of Oral-Digital Habits."

72 A study of children: "Situation-Response (S-R) Questions for Identifying the Function of Problem Behaviour: The Example of Thumb Sucking," *British Journal of Clinical Psychology* 29, no. 1 (1990).

5. PACIFIERS: THE NEXT MENACE

75 A physician wrote: O. Tonz, "Breastfeeding in Modern and Ancient Times: Facts, Ideas and Beliefs," in *Short and Long Term Effects of Breast Feeding on Child Health*, 6.

77 But a recent study, using: Gale and Martyn, "Dummies and the Health of Hertfordshire Infants."

77 And Spock had: Benjamin Spock, *The Common Sense Book of Baby and Child Care*, 2nd ed. (New York: Duell, Sloan, and Pearce, 1957), 215.

77 The question "Has he": John and Elizabeth Newson, *Infant Care in an Urban Community* (New York: International Universities Press, 1963), 66.

77 On the other side: Ibid., 67

78 "I mean, when": Ibid., 67

78 Few people say: J. Whitmarsh, "Mums, Dummies and 'Dirty Dids': The

Dummy as a Symbolic Representation of Mothering?" *Children & Society* 22, no. 4 (2008).

78 The core assumption: J. Whitmarsh, "The Good, the Bad and the Pacifier: Unsettling Accounts of Early Years Practice," *Journal of Early Childhood Research* 6, no. 2 (2008): 148.

80 The Baby-Friendly: www.babyfriendlyusa.org (3 October 2012).

80 In fact, the best studies: For a review, See N. R. O'Connor et al., "Pacifiers and Breastfeeding: A Systematic Review," *Archives of Pediatrics & Adolescent Medicine* 163, no. 4 (2009). Also see A. G. Jenik and N. Vain, "The Pacifier Debate," *Early Human Development* 85, no. 10 suppl. (2009).

80 In fact, a recent study: "Increase in Supplemental Formula Feeds Observed Following Removal of Pacifiers from a Mother Baby," www .abstracts2view.com/pas/view.php?nu=PAS12L1_1090&terms (accessed 3 October 2012).

80 Nipple confusion: D. Dowling and W. Thanattherakul, "Nipple Confusion, Alternative Feeding Methods, and Breast-Feeding Supplementation: State of the Science," *Newborn and Infant Nursing Reviews* 1, no. 4 (2001). Also see C. Fisher and S. Inch, "Nipple Confusion—Who Is Confused?" *Journal of Pediatrics* 129, no. 1 (1996).

80 The current edition: Jan Riordan and Karen Wambach, *Breastfeeding and Human Lactation*, 4th ed. (Sudbury, MA: Jones and Bartlett, 2010).

81 There is evidence: J. Whitmarsh, "The Good, the Bad and the Pacifier."

81 Pacifiers are highly: E. Blass, "Suckling," *Encyclopedia of Infant and Early Childhood Development* (2008).

81 They also seem to promote: T. Field, "Sucking and Massage Therapy Reduce Stress during Infancy," in *Soothing and Stress,* eds. M. Lewis and D. Ramsay (Mahwah, NJ: Lawrence Erlbaum, 1999).

82 A series of studies: Ibid.

82 A recent book: Tiffany Field, *The Amazing Infant* (Malden, MA: Blackwell, 2007), 253.

82 It's as if: See Kukla's acute analysis in *Mass Hysteria*.

83 This is why: Blass, "Suckling," *Encyclopedia of Infant and Early Childhood Development,* 283.

83 As the anthropologist: Hrdy, *Mother Nature,* 388.

PART TWO: SMILE

87 "What you doing?": Kenneth Kaye, *The Mental and Social Life of Babies: How Parents Create Persons* (Chicago: University of Chicago Press, 1982), 192.

6. THE BIRTH OF A SMILE

Here are several: Ann Oakley, *Becoming a Mother* (Oxford: Martin Robertson, 1979). Quotations (in order) on 115, 91, 97, 117, 116.

92 "After such hard labor": Ibid., 116.

94 "Parents observing for": Philippe Rochat, *The Infant's World* (Cambridge, MA: Harvard University Press, 2001), 183.

94 The moment when: Daniel Stern, *The Interpersonal World of the Infant: A View from Psychoanalysis and Developmental Psychology* (New York: Basic Books, 1985), 37.

94 The Anbarra: Annette Hamilton, *Nature and Nurture: Aboriginal Child-Rearing in North-Central Arnhem Land* (Canberra: Australian Institute of Aboriginal Studies, 1981).

95 When people go: R. E. Kraut and R. E. Johnston, "Social and Emotional Messages of Smiling: An Ethological Approach," *Journal of Personality and Social Psychology* 37 (1979).

95 A similar study: D. Matsumoto and B. Willingham, "The Thrill of Victory and the Agony of Defeat: Spontaneous Expressions of Medal Winners of the 2004 Athens Olympic Games," *Journal of Personality and Social Psychology* 91, no. 3 (2006).

95 Even *implied* social: A. Fridlund, "Sociality of Solitary Smiling: Potentiation by an Implicit Audience," *Journal of Personality and Social Psychology* 60, no. 2 (1991).

95 "There is no point": Vasudevi Reddy, *How Infants Know Minds* (Cambridge, MA: Harvard University Press, 2008), 82.

95 To quote Melville: Herman Melville, *Pierre; or, The Ambiguities* (1852; reprint, New York: Penguin, 1996), 84.

96 Some cultures value smiling: H. Keller and H. Otto, "The Cultural Socialization of Emotion Regulation During Infancy," *Journal of Cross-Cultural Psychology* 40 (2009).

98 Only nine minutes: C. C. Goren, M. Sarty, P. Y. Wu, "Visual Following and Pattern Discrimination of Face-like Stimuli by Newborn Infants," *Pediatrics* 56, no. 4 (1975).

98 Newborns only a few: A. Slater and P. C. Quinn, "Face Recognition in the Newborn Infant," *Infant and Child Development* 10, nos. 1–2 (2001).

98 Because most infants: P. C. Quinn et al., "Representation of the Gender of Human Faces by Infants: A Preference for Female," *Perception* 31 (2002).

99 But as they approach: O. Pascalis, M. Haan, and C. A. Nelson, "Is Face Processing Species-Specific During the First Year of Life?" *Science* 296, no. 5571 (2002).

100 In a paper that's been: A. N. Meltzoff and M. K. Moore, "Imitation of Facial and Manual Gestures by Human Neonates," *Science* 198, no. 4312 (1977).

100 Meltzoff arranged: A. Gopnik, A. N. Meltzoff, and P. K. Kuhl, *The Scientist in the Crib: What Early Learning Tells Us About the Mind* (New York: Harper-Collins, 2000).

100 A six-week-old: A. N. Meltzoff and M. K. Moore, "Imitation, Memory, and the Representation of Persons," *Infant Behavior and Development* 17, no. 1

(1994). Also see A. N. Meltzoff and M. K. Moore, "Early Imitation Within a Functional Framework: The Importance of Person Identity, Movement, and Development," *Infant Behavior and Development* 15, no. 4 (1992).

101 The scientists were studying: A popular account of the discovery and its implications is Marco Iacoboni's *Mirroring People: The New Science of How We Connect with Others* (New York: Farrar, Straus and Giroux, 2008).

101 "The child, even": A. N. Meltzoff, "Roots of Social Cognition: The *Like-Me* Framework," in *Minnesota Symposia on Child Psychology: Meeting the Challenge of Translational Research in Child Psychology* 35, eds. D. Cicchetti and M. R. Gunnar (Hoboken, NJ: Wiley, 2009), 35.

7. "A MOTHER EVIDENTLY PERCEIVES HER BABY TO BE A PERSON LIKE HERSELF"

103 For much of the twentieth: A. N. Meltzoff and W. Prinz, eds., *The Imitative Mind: Development, Evolution and Brain Bases* (New York: Cambridge University Press, 2002), 19.

105 Babies treat objects: C. Trevarthen, "Communication and Cooperation in Early Infancy: A Description of Primary Intersubjectivity," in *Before Speech: The Beginning of Interpersonal Communication,* ed. M. Bullowa (New York: Cambridge University Press, 1979), 323.

105 He can develop: Ibid., 321.

105 At only a couple: Ibid., 347.

106 After his first son: Darwin, "A Biographical Sketch of an Infant," 293–94.

106 The old ideas: For a personal account of this early rush of research from Trevarthen and others, see "The Concepts and Foundations of Infant Intersubjectivity," in *Intersubjective Communication and Emotion in Early Ontogeny,* ed. Stein Bråten (New York: Cambridge University Press, 1998).

106 "These interactions": M. C. Bateson, " 'The Epigenesis of Conversational Interaction': A Personal Account of Research Development," in *Before Speech,* ed. Bullowa, 65.

107 A brilliant: L. Murray and C. Trevarthen, "Emotional Regulation of Interactions Between Two-Month-Olds and Their Mothers," in *Social Perception in Infants,* eds. T. Field and N. Fox (Norwood, NJ: Ablex, 1985). This study remains controversial. For contrasting perspectives on its worth, see P. Rochat, U. Neisser, and V. Marian, "Are Young Infants Sensitive to Interpersonal Contingency?" *Infant Behavior and Development* 21, no. 2 (1998); and J. Nadel et al., "Expectancies for Social Contingency in 2-Month-Olds," *Developmental Science* 2, no. 2 (1999).

108 Edward Tronick: E. Z. Tronick, H. Als, and L. Adamson, "Structure of Early Face-to-Face Communicative Interactions," in *Before Speech,* ed. Bullowa, 369.

108 These dialogues: P. Rochat and T. Striano, "Social-Cognitive Development in the First Year," in *Early Social Cognition: Understanding Others in*

the First Months of Life, ed. P. Rochat (Mahwah, NJ: Lawrence Erlbaum, 1999), 9.

108 "A mother evidently": Trevarthen, "Communication and Cooperation in Early Infancy," 340.

109 "Until recently most": M. Bullowa, "Prelinguistic Communication: A Field for Scientific Research," in *Before Speech*, 1, 31.

110 "It was not so long": Reddy, *How Infants Know Minds*, 5–6.

113 Babies who are blind: S. Fraiberg, "Blind Infants and Their Mothers: An Examination of the Sign System," in *The Effect of the Infant on Its Caregiver*, eds. M. Lewis and L. Rosenblum (Oxford: Wiley, 1974).

113 Detailed studies: M. Lavelli and A. Fogel, "Developmental Changes in Mother-Infant Face-to-Face Communication: Birth to 3 Months," *Developmental Psychology* 38, no. 2 (2002).

113 These early face-to-face: P. Rochat, "Intentional Action Arises from Early Reciprocal Exchanges," *Acta Psychologica* 124 (2007): 21.

114 When Tomkins began: Malcolm Gladwell, "The Naked Face," *New Yorker*, August 5, 2002.

115 The bared-teeth display: B. M. Walker and R. M. Dunbar, "Differential Behavioural Effects of Silent Bared Teeth Display and Relaxed Open Mouth Display in Chimpanzees," *Ethology* 111 (2005).

115 This connection was: For a popular account of the argument over "higher" emotions and of Ekman's odyssey, see "Darwin's Joys" in Dacher Keltner's *Born to Be Good* (New York: W. W. Norton, 2009). Ekman's own internecine recounting is his afterword to Charles Darwin, *The Expression of the Emotions in Man and Animals* (Oxford: Oxford University Press, 1998).

115 Darwin argued: Darwin, *The Expression of the Emotions in Man and Animals* (1998), 348.

8. GET HAPPY

118 The human face: For more information on FACS, including links to chapters of the FACS manual, see www.face-and-emotion.com/dataface/facs/description.jsp (3 October 2012). The quotation is taken from: www.face-and-emotion.com/dataface/facs/manual/AU1.html (3 October 2012).

118 Darwin, recalling his: Darwin, *The Expression of the Emotions*, 209.

119 Duchenne's book: Duchenne de Boulogne, *The Mechanism of Human Facial Expression* (New York: Cambridge University Press, 1990).

119 The orbicularis oculi, Duchenne: Ibid., 72.

120 Paul Ekman has: Paul Ekman, *Telling Lies: Clues to Deceit in the Marketplace, Politics, and Marriage*, 3rd ed. (New York: W. W. Norton, 2009).

120 Show an American: From an unpublished 1972 dissertation by Wallace Friesen; described in D. Matsumoto and C. Kupperbusch, "Idiocentric and Allocentric Differences in Emotional Expression, Experience, and the

Coherence Between Expression and Experience," *Asian Journal of Social Psychology* 4, no. 2 (2001).

121 There is emoticon-based: M. Yuki, W. W. Maddux, and T. Masuda, "Are the Windows to the Soul the Same in the East and West?" *Journal of Experimental Social Psychology* 43 (2007).

122 According to a recent article: M. J. Hertenstein, "Cautions in the Study of Infant Emotional Displays," *Emotion Review* 2, no. 2 (2010): 130.

122 "The impetus for": Sarah Blaffer Hrdy, *Mothers and Others: The Evolutionary Origins of Mutual Understanding* (Cambridge, MA: Harvard University Press, 2009), 38.

123 "Part of the modern": P. T. Ellison, "A Review of Sarah Blaffer Hrdy, *Mothers and Others*," *Evolutionary Psychology* 7, no. 3 (2009): 443.

123 Humans engage with: Hrdy, *Mothers and Others*, 117.

124 "We do not see": Peter Hobson, *The Cradle of Thought* (London: Macmillan, 2002), 243.

125 At the very least, there's: See Seth Pollak's summary in (2009): 37. "The Emergence of Emotion: Experience, Development, and Biology," in *Minnesota Symposia on Child Psychology* 35.

125 Between four and twelve months: D. S. Bennett, M. Bendersky, and M. Lewis, "Does the Organization of Emotional Expression Change Over Time? Facial Expressivity from 4 to 12 Months," *Infancy* 8, no. 2 (2005).

126 The power of this: J. F. Sorce et al., "Maternal Emotional Signaling: Its Effect on the Visual Cliff Behavior of 1-Year-Olds," *Developmental Psychology* 21, no. 1 (1985).

127 At this early age: M. Lavelli and A. Fogel, "Developmental Changes in the Relationship Between the Infant's Attention and Emotion During Early Face-to-Face Communication: The 2-Month Transition," *Developmental Psychology* 41, no. 1 (2005).

127 The smile plays: For a succinct, accessible overview of the science of social smiling, see Daniel Messinger's "Smiling" in *Encyclopedia of Infant and Early Childhood Development* (2008).

128 The practical effect: D. S. Messinger and A. Fogel, "The Interactive Development of Social Smiling," *Advances in Child Development and Behavior* 35 (2007): 348.

128 This is what's been fired: J. F. Cohn and E. Z. Tronick, "Mother-Infant Face-to-Face Interaction: The Sequence of Dyadic States at 3, 6, and 9 months," *Developmental Psychology* 23 (1987).

128 Working with his advisor: D. Messinger, A. Fogel, and K. L. Dickson, "All Smiles Are Positive, but Some Smiles Are More Positive than Others," *Developmental Psychology* 37, no. 5 (2001).

129 In fact, they seem: A. Fogel et al., "Effects of Normal and Perturbed Social

Play on the Duration and Amplitude of Different Types of Infant Smiles," *Developmental Psychology* 42, no. 3 (2006): 469.

131 Sometimes you can: Messinger, Fogel, and Dickson, "All Smiles Are Positive."

132 When first-time mothers: L. Strathearn et al., "What's in a Smile? Maternal Brain Responses to Infant Facial Cues," *Pediatrics* 122, no. 1 (2008).

133 When depressed mothers: M. Bornstein et al., "Discrimination of Facial Expression by 5-Month-Old Infants of Nondepressed and Clinically Depressed Mothers," *Infant Behavior and Development* 34, no. 1 (2011): 101.

133 "From the American": A. G. Halberstadt and F. T. Lozada, "Emotion Development in Infancy through the Lens of Culture," *Emotion Review* 3, no. 2 (2011): 163.

134 In ethnographic: J. E. Kilbride and P. L. Kilbride, "Sitting and Smiling Behavior of Baganda Infants: The Influence of Culturally Constituted Experience," *Journal of Cross-Cultural Psychology* 6 (1975): 93–94.

136 "Given these limitations": Ed Tronick, *The Neurobehavioral and Social-Emotional Development of Infants and Children* (New York: W. W. Norton, 2007), 166.

9. A FEW WORDS ON SOME SMALL SUBJECTS LIKE CULTURE, CIVILIZATION, AND THE ORIGINS OF HAPPINESS

138 Humans aren't the: For a review of social gaze among different species, see N. J. Emery, "The Eyes Have It: The Neuroethology, Function and Evolution of Social Gaze," *Neuroscience and Biobehavioral Reviews* 24 (2000).

139 Without this sort: M. Tomasello et al., "Reliance on Head Versus Eyes in the Gaze Following of Great Apes and Human Infants: The Cooperative Eye Hypothesis," *Journal of Human Evolution* 52, no. 3 (2007). Also see Michael Tomasello, "For Human Eyes Only," *New York Times*, January 13, 2007.

139 Ideally, as the psychologist: P. Hobson, "Who Puts the Jointness into Joint Attention?" in *Joint Attention: Communication and Other Minds*, eds. N. Eilan et al. (New York: Oxford University Press, 2005), 185.

139 At around a year: M. Tomasello, "Joint Attention and Social Cognition," in *Joint Attention: Its Origins and Role in Development*, eds. C. Moore and P. J. Dunham (Hillsdale, NJ: Lawrence Erlbaum, 1995), 104.

140 Evolutionarily, it is fundamental: M. Tomasello and H. Moll, "The Gap Is Social: Human Shared Intentionality and Culture," in *Mind the Gap: Tracing the Origins of Human Universals*, eds. J. Silk and P. M. Kappeler (New York: Springer, 2009).

140 It's that we were: Ibid., 332.

140 When babies and adults: M. Tomasello et al., "Understanding and Sharing Intentions: The Origins of Cultural Cognition," *Behaviorial and Brain Sciences* 28 (2005): 683.

141 Instead, they appear: Messinger and Fogel, "The Interactive Development of Social Smiling."

141 Twenty years ago: S. S. Jones, K. Collins, and H.-W. Hong, "An Audience Effect on Smile Production in 10-Month-Old Infants," *Psychological Science* 2, no. 1 (1991): 45.

142 But it turned out: Ibid., 48.

143 There's growing evidence: S. S. Jones and H.-W. Hong, "Onset of Voluntary Communication: Smiling Looks to Mother," *Infancy* 2, no. 3 (2001).

143 Unlike these reactive: M. Venezia et al., "The Development of Anticipatory Smiling," *Infancy* 6, no. 3 (2004).

144 And somewhat amazingly: M. V. Parlade et al., "Anticipatory Smiling: Linking Early Affective Communication and Social Outcome," *Infant Behavior and Development* 32, no. 1 (2009).

PART THREE: TOUCH

147 "10. Handle infant": Barbara Rogoff, *The Cultural Nature of Human Development* (New York: Oxford University Press, 2003), 130.

148 "And I kept": Stephen Humphries and Pamela Gordon, *A Labour of Love: The Experience of Parenthood in Britain, 1900–1950* (London: Sidgwick & Jackson, 1993), 69.

148 "On a trip to": Nina Jablonski, *Skin: A Natural History* (Berkeley: University of California Press, 2006), 206.

10. THE POWER OF TOUCH

149 "The question then": S. M. Schanberg and T. M. Field, "Sensory Deprivation Stress and Supplemental Stimulation in the Rat Pup and Preterm Human Neonate," *Child Development* 58, no. 6 (1987): 1434. The story of Schanberg's discovery is told in Tiffany Field's *Touch* (Cambridge, MA: MIT Press, 2001).

152 A classic experiment: M. Kaitz et al., "Parturient Women Can Recognize Their Infants by Touch," *Developmental Psychology* 28, no. 1 (1992).

152 In studies of: Jablonski, *Skin*, 107.

152 Another experiment, this time: E. Z. Tronick, "Touch in Mother-Infant Interaction," in *Touch in Early Development*, ed. T. M. Field (Mahwah, NJ: Lawrence Erlbaum, 1995).

152 Newborns who are held: L. Gray, L. Watt, and E. M. Blass, "Skin-to-Skin Contact Is Analgesic in Healthy Newborns," *Pediatrics* 105, no. 1 (2000).

152 It is, he wrote: Robert Jütte, *A History of the Senses: From Antiquity to Cyberspace* (Cambridge, UK: Polity, 2005), 70.

153 "Touch is at the": Sander Gilman, *Inscribing the Other* (Lincoln: University of Nebraska Press, 1991), 31.

153 "The physiology of touch": Ibid.

154 "Touch is ten": Diane Ackerman, *A Natural History of the Senses* (New York: Random House, 1990), 77–78.

155 But between two and six: L. B. Jahromi, S. P. Putnam, and C. A. Stifter, "Maternal Regulation of Infant Reactivity from 2 to 6 Months," *Developmental Psychology* 40, no. 4 (2004).

155 Touch has what: For more see M. J. Hertenstein, "Touch: Its Communicative Functions in Infancy," *Human Development* 45, no. 2 (2002).

156 Generosity of touch: S. Suomi, "Touch and the Immune System in Rhesus Monkeys," in *Touch in Early Development*, ed. T. M. Field.

156 In some primate: Jablonski, *Skin*, 108.

156 We have empathy: The existence of empathy in different species is a hotly contested topic; for a popular account of the argument against, see Hrdy's *Mothers and Others*, focusing on the work of Joan Silk.

157 It's just much, much: Jablonski, *Skin*, 19.

157 At this late point: See Dean Falk's explanation of our fate in *Finding Our Tongues: Mothers, Infants, and the Origins of Language* (New York: Basic Books, 2010).

157 "In a sense": Ibid., 44.

158 "For more than": Hrdy, *Mother Nature*, 97.

158 "Would any biped": C. M. Wall-Scheffler, K. Geiger, and K. L. Steudel-Numbers, "Infant Carrying: The Role of Increased Locomotory Costs in Early Tool Development," *American Journal of Physical Anthropology* 133, no. 2 (2007): 845.

11. A BRIEF HISTORY OF HOW WE HAVE, AND HAVE NOT, HELD OUR CHILDREN

163 When a child: Tronick, *The Neurobehavioral and Social-Emotional Development of Infants and Children*.

164 The seventeenth-century: Karin Calvert, *Children in the House : The Material Culture of Early Childhood, 1600–1900* (Boston: Northeastern University Press, 1992), 19.

164 In other words: Ethnographic accounts of "the first touch" are taken from Nancy Caldwell Sorel, ed., *Ever Since Eve: Personal Reflections of Childbirth* (New York: Oxford University Press, 1984); David Meltzer, ed., *Birth* (New York: Ballantine Books, 1973); Margarita Artschwager Kay, ed., *Anthropology of Human Birth* (Philadelphia: F. A. Davis, 1982); and the exhaustive *Encyclopedia of Medical Anthropology*, ed. C. Ember and M. Ember (New York: Springer, 2003).

166 In the early 1800s: Calvert, *Children in the House*, 62.

167 Even today: Eugenia Georges, *Bodies of Knowledge: The Medicalization of Reproduction in Greece* (Nashville: Vanderbilt University Press, 2008). The practice, although dangerous, is not rare: see the impassioned presentation, "Salting Newborns: Pickling Them or Killing Them? A Practice That

Should Be Stopped," medical.abu-osba.com/PublishedPapers/20091514331 .ppt (3 October 2012).

168 The limbs of: Lawrence Stone, *The Family, Sex and Marriage in England, 1500–1800* (New York: Harper & Row, 1977), 162.

168 "In the swaddling": Jane Sharp, *The Midwives Book: Or the Whole Art of Midwifry Discovered* (1671; reprint, New York: Oxford University Press, 1999), 272.

168 Straight limbs: E. L. Lipton, A. Steinschneider, and J. B. Richmond, "Swaddling, a Child Care Practice: Historical, Cultural, and Experimental Observations," *Pediatrics* 35, no. 3 (1965): 523.

168 All this is why: Simon Schama, *The Embarrassment of Riches: An Interpretation of Dutch Culture in the Golden Age* (New York: Knopf, 1987), 537.

169 "The infant is bound": Jean-Jacques Rousseau, *Emile* (New York: Dutton, 1966), 11–12.

169 "The truth is": Lipton, Steinschneider, and Richmond, "Swaddling, a Child Care Practice," 527–28.

169 Rousseau himself: Rousseau, *Emile*, 27.

170 In 1870: *The Bazar Book of Decorum* (New York: Harper & Brothers, 1870), 78.

170 By the early 1950s: Lipton, Steinschneider, and Richmond, "Swaddling, a Child Care Practice," 563.

170 For the infant: Geoffrey Gorer, *The People of Great Russia: A Psychological Study* (London: Cresset Press, 1949), 123.

170 The anthropologist Margaret: M. Mead, "National Character," in *Anthropology Today: An Encyclopedic Inventory*, ed. A. L. Kroeber (Chicago: University of Chicago Press, 1953), 644.

171 "Could it be": Lipton, Steinschneider, and Richmond, "Swaddling, a Child Care Practice," 521.

171 "It is conceivable": Ibid., 564.

171 And it did: B. E. van Sleuwen, "Swaddling: A Systematic Review," *Pediatrics* 120, no. 4 (2007): 1098.

172 The studies confirm: Ibid., for a thorough review of the science.

172 The most detailed study: J. S. Chisholm, "Swaddling, Cradleboards, and the Development of Children," *Early Human Development* 2, no. 3 (1978). For more detail, see Chisholm's *Navajo Infancy: An Ethological Study of Child Development* (Hawthorne, NY: Aldine, 1983).

173 "In spite of": Ibid., 270.

173 A recent randomized trial: S. Manaseki-Holland et al., "Effects of Traditional Swaddling on Development: A Randomized Controlled Trial," *Pediatrics* 126, no. 6 (2010).

175 Orphanages, then called: Robert Sapolsky, *Monkeyluv: And Other Essays on Our Lives as Animals* (New York: Scribner, 2005), 152.

175 The condition was: H. Bakwin, "Emotional Deprivation in Infants," *The Journal of Pediatrics* 35, no. 4 (1949): 512, 514.

176 At Bellevue: Robert Karen, *Becoming Attached: First Relationships and How They Shape Our Capacity to Love* (New York: Warner Books, 1994).

176 Descartes famously: Anthony Synnott, "Handling Children: To Touch or Not to Touch?" in *The Book of Touch*, ed. C. Classen (New York: Berg, 2005), 43.

176 "Are there any": Luther Emmett Holt, *The Care and Feeding of Children* (New York: D. Appleton & Co., 1918), 175.

177 "Never hug and kiss": Watson, *Psychological Care of Infant and Child*, 81–82.

177 Watson argued that: Ibid., 71, 75.

178 Benjamin Spock: Benjamin Spock, *The Common Sense Book of Baby and Child Care* (New York: Duell, Sloan and Pearce, 1946), 19.

179 "Indeed, the disparity": H. Harlow, "The Nature of Love," *American Psychologist* 13 (1958): 677.

180 The deep connection: John Bowlby, *Child Care and the Growth of Love* (Harmondsworth, UK: Penguin, 1953), 240.

181 In the decades: For the history of attachment theory, see Karen, *Becoming Attached*.

12. A BLESSED BREAK FROM EMOTION AND A SUDDEN DETOUR INTO PHYSIOLOGY, CARNIVAL MIDWAYS, AND KANGAROOS

183 Flayed, an adult's skin: Field, *Touch*.

183 "If you think": Ackerman, *A Natural History of the Senses*, 68.

184 The most Fragite: G. L. Darmstadt and J. G. Dinulos, "Neonatal Skin Care," *Pediatric Clinics of North America* 47, no. 4 (2000).

185 Harry Chapin: Jeffrey P. Baker, "Technology in the Nursery," in *Formative Years: Children's Health in the United States, 1880–2000*, eds. A. M. Stern and H. Markel (Ann Arbor: University of Michigan Press, 2002), 73.

185 The carnival displays: The weird story of Martin Couney is told by Deborah Blum in *Love at Goon Park: Harry Harlow and the Science of Affection* (Cambridge, MA: Perseus, 2002) and (more skeptically) by Jeffrey Baker in *The Machine in the Nursery: Incubator Technology and the Origins of Newborn Intensive Care* (Baltimore: Johns Hopkins University Press, 1996). Also see A. J. Liebling's comic "A Patron of the Preemies," *New Yorker*, June 3, 1959.

186 In the first incubator: Murdina M. Desmond, *Newborn Medicine and Society: European Background and American Practice, 1750–1975* (Austin, TX: Eakin Press, 1998), 75.

186 "When I was a pediatric": Alistair Philip, "The Evolution of Neonatology," *Pediatric Research* 58, no. 4 (2005): 808.

186 A study of a NICU: A. W. Gottfried and J. L. Gaiter, eds., *Infant Stress Under Intensive Care: Environmental Neonatology* (Baltimore: University Park Press, 1985), 49.

187 But we should not: "Mothers of Premature Babies," *British Medical Journal* 2 (1970): 556.

187 An experiment in: C. R. Barnett et al., "Neonatal Separation: The Maternal Side of Interactional Deprivation," *Pediatrics* 45, no. 2 (1970).

188 The San Juan: See Nathalie Charpak's *Kangaroo Babies: A Different Way of Mothering* (London: Souvenir Press, 2006); and Susan Ludington-Hoe's *Kangaroo Care: The Best You Can Do to Help Your Preterm Infant* (New York: Bantam, 1993).

189 An example: being: R. Feldman et al., "Skin-to-Skin Contact (Kangaroo Care) Promotes Self-Regulation in Premature Infants: Sleep-Wake Cyclicity, Arousal Modulation, and Sustained Exploration," *Developmental Psychology* 38, no. 2 (2002).

190 A recent review of the technique: N. Charpak et al., "Kangaroo Mother Care: 25 Years After," *Acta Paediatrica* 94 (2005): 515. In general, though, this is a solid review of the science.

190 The first medical: A. Whitelaw and K. Sleath, "Myth of the Marsupial Mother: Home Care of Very Low Birth Weight Babies in Bogota, Columbia," *The Lancet* 325, no. 8439 (1985): 1208.

190 Their mothers are: R. Feldman et al., "Comparison of Skin-to-Skin (Kangaroo) and Traditional Care: Parenting Outcomes and Preterm Infant Development," *Pediatrics* 110, no. 1 (2002).

191 Even for infants born: S. G. Ferber and I. R. Makhoul, "The Effect of Skin-to-Skin Contact (Kangaroo Care) Shortly After Birth on the Neurobehavioral Responses of the Term Newborn: A Randomized, Controlled Trial," *Pediatrics* 113, no. 4 (2004).

191 The evidence for kangaroo: J. E. Lawn et al., " 'Kangaroo Mother Care' to Prevent Neonatal Deaths Due to Preterm Birth Complications," *International Journal of Epidemiology* 39, suppl. 1 (2010).

193 And in a seminal study: T. Field et al., "Tactile/Kinesthetic Stimulation Effects on Preform Neonates," *Pediatrics* 77, no. 5 (1986).

13. IN WHICH TOUCH GETS PERHAPS A LITTLE TOO MUCH POWER

196 Proximity itself: Kukla, *Mass Hysteria*, 148.

196 Sarah Hrdy has: Hrdy, *Mothers and Others*, 493.

198 Robert Karen: Karen, *Becoming Attached*, 70.

199 "What is wrong": Ibid., 78.

199 "I was angry": Ibid., 81.

200 "If they decided": Ibid., 85.

200 In their 1976: Marshall H. Klaus and John H. Kennell, *Maternal-Infant Bonding: The Impact of Early Separation or Loss on Family Development* (Saint Louis: Mosby, 1976), 50.

201 As Diane Eyer: Diane E. Eyer, *Mother-Infant Bonding: A Scientific Fiction* (New Haven: Yale University Press, 1992), 20. Eyer's book is a merciless deconstruction of the evidence for bonding.

201 Here's the well-known: Christina Hardyment, *Dream Babies: Childcare Advice from John Locke to Gina Ford* (London: Frances Lincoln, 2007), 317.

202 A headline: "The One Day Old Deprived Child," *New Scientist*, March 28, 1974.

202 "We're all getting": Marshall H. Klaus and John H. Kennell, *Parent-Infant Bonding* (Saint Louis: Mosby, 1982), 56.

202 Except that bonding did: Hrdy, *Mother Nature*, 487–88.

202 The zoologist: P. H. Klopfer, "'Mother Love' Revisited: On the Use of Animal Models," *American Scientist* 84, no. 4 (1996): 319.

203 On the Dr. Sears: www.askdrsears.com/topics/pregnancy-childbirth/tenth-month-post-partum/bonding-your-newborn/bonding-what-it-means (3 October 2012).

203 In the most recent: William Sears et al., *The Portable Pediatrician: Everything You Need to Know About Your Child's Health* (New York: Little, Brown, 2011), 45–46.

203 "Catch-up bonding": www.askdrsears.com/topics/pregnancy-childbirth/tenth-month-post-partum/bonding-your-newborn/bonding-what-it-means (3 October 2012).

204 A recent academic: R. Feldman, "Maternal-Infant Contact and Child Development: Insights from the Kangaroo Intervention," in *Low-Cost Approaches to Promote Physical and Mental Health* (New York: Springer, 2007), 324, 331.

205 He was omnivorous: Ashley Montagu, *Touching: The Human Significance of the Skin* (New York: HarperCollins, 1986). Quotations are from, in order: 47, 180, 35.

205 Large sections of the book: Ibid., 157.

205 "Can you see": Sharon Heller, *The Vital Touch: How Intimate Contact with Your Baby Leads to Happier, Healthier Development* (New York: Holt, 1997), 49.

207 The philosopher: David Hume, "The Sceptic," in *Selected Essays* (New York: Oxford University Press, 1998), 95.

209 The unique: M. Konner, "Maternal Care, Infant Behavior and Development among the !Kung," *Kalahari Hunter-Gatherers: Studies of the !Kung San and Their Neighbors*, eds. R. B. Lee and I. DeVore (Cambridge, MA: Harvard University Press, 1976), 219.

209 Infants in industrialized: Barry S. Hewlett, "Diverse Contexts of Human Infancy," in *Cross-Cultural Research for Social Science*, eds. C. Ember and M. Ember (Englewood Cliffs, NJ: Prentice Hall, 1996).

209 If you have a vague: M. Konner, "Aspects of the Developmental Ethology of a Foraging People," in *Ethological Studies of Child Behavior*, ed. N. G. B. Jones (New York: Cambridge University Press, 1972), 292.

210 And if she did: Jean Liedloff, *The Continuum Concept: In Search of Happiness Lost* (New York: Da Capo Press, 1986), 71.

211 Sample sentence: Ibid., 62, 69.

211 For Liedloff: Ibid., 163, 162.

211 If a mother carries: Ibid., 160.

212 It remains: Chris Bobel, *The Paradox of Natural Mothering* (Philadelphia: Temple University Press, 2002), 88.

212 "Liedloff's work earned": C. Bobel, "When Good Isn't Enough: Mother Blame in *The Continuum Concept*," *Journal of the Association for Research on Mothering* 6, no. 2 (2004): 68

212 "What's their secret?": Harvey Karp, *The Happiest Baby on the Block* (New York: Random House, 2003).

213 An Australian woman: www.attachmentparentingaustralia.com/experiences.htm (3 October 2012).

214 In fact, the "rightness": U. A. Hunziker and R. G. Barr, "Increased Carrying Reduces Infant Crying: A Randomized Controlled Trial," *Pediatrics* 77, no. 5 (1986).

214 As Konner observed: M. Konner, *Kalahari Hunter-Gatherers*, 220.

214 But despite this: M. Konner, "The Hunter-Gatherer Infancy and Childhood: The !Kung and Others," in *Hunter-Gatherer Childhoods: Evolutionary, Developmental, and Cultural Perspectives*, eds. B. S. Hewlett and M. E. Lamb (New Brunswick, NJ: Aldine de Gruyter, 2005), 23.

216 Alloparenting is a cushion: Ibid., 107.

216 "Given frequent changes": Alma Gottlieb, *The Afterlife Is Where We Come From: The Culture of Infancy in West Africa* (Chicago: University of Chicago Press, 2004), 140–41.

216 "Biology is no more": Tronick, *The Neurobehavioral and Social-Emotional Development of Infants and Children*, 122.

216 Instead, he says: Tronick, "Touch in Mother-Infant Interaction," 58–59.

217 "Our decisions about child": Tronick, *The Neurobehavioral and Social-Emotional Development of Infants and Children*, 103.

217 Margaret Mead's hope: See the first chapter of Robert A. Levine et al., *Child Care and Culture: Lessons from Africa* (New York: Cambridge University Press, 1994).

218 Wowed by: Blum, *Love at Goon Park*, 193.

218 Harlow had revolutionized: Ibid.

218 The psychologist Jerome: Jerome Kagan, *The Nature of the Child* (New York: Basic Books, 1994), 55.

219 As the neuroscientist Robert: Robert M. Sapolsky, *Why Zebras Don't Get Ulcers: A Guide to Stress, Stress-Related Diseases, and Coping* (New York: W. H. Freeman, 1994), 109.

14. FINDING THE NEW WORLD

221 when the psychologist Philippe: Rochat, *The Infant's World*, 54.

223 According to the psychologist Claes: C. von Hofsten, "An Action Perspective on Motor Development," *Trends in Cognitive Sciences* 8, no. 6 (2004): 267.

224 The development of independent: B. Bertenthal and C. von Hofsten, "Eye, Head and Trunk Control: The Foundation for Manual Development," *Neuroscience and Biobehavioral Reviews* 22, no. 4 (1998): 517.

225 A few years ago, some: J. A. Sommerville, A. L. Woodward, and A. Needham, "Action Experience Alters 3-Month-Old Infants' Perception of Others' Actions," *Cognition* 96, no. 1 (2005).

PART FOUR: TODDLE

227 "Learning to walk": Calvert, *Children in the House*, 36–37.

227 "I can't remember": Esther Thelen, "The Improvising Infant: Learning About Learning to Move," in *The Developmental Psychologists: Research Adventures Across the Life Span*, eds. M. R. Merrens and G. G. Brannigan (New York: McGraw-Hill, 1996), 39

15. WHO PUT THE NORM IN NORMAL?

230 In her history: Ann Hulbert, *Raising America: Experts, Parents, and a Century of Advice About Children* (New York: Knopf, 2003), 175.

232 Measurement was a novel: André Turmel, *A Historical Sociology of Childhood: Developmental Thinking, Categorization, and Graphic Visualization* (New York: Cambridge University Press, 2008), 86.

232 That's why he: Ibid., 93–94.

233 "Cinema analysis": Arnold Gesell and Helen Thompson, *Infant Behavior: Its Genesis and Growth* (New York: McGraw-Hill, 1934), 19.

233 In two bullet-stopping: Arnold Gesell, *An Atlas of Infant Behavior: A Systematic Delineation of the Forms and Early Growth of Human Behavior Patterns, Normative Series* (New Haven: Yale University Press, 1934), 41.

233 From the *Atlas*: Ibid., 241, 371.

234 "We think of behavior": Gesell and Thompson, *Infant Behavior*, 61.

234 "It is probably hard": E. Thelen and K. E. Adolph, "Arnold L. Gesell: The Paradox of Nature and Nurture," *Developmental Psychology* 28, no. 3 (1992): 374. A remarkably insightful essay on Gesell, written from a developmental perspective.

234 A new consciousness: See Howard P. Chudacoff, *How Old Are You?: Age Consciousness in American Culture* (Princeton: Princeton University Press, 1992).

235 For Gesell, mental "normalness": Turmel, *A Historical Sociology of Childhood*, 232–33, 274.

236 From *Infant and Child*: Thelen and Adolph, "Arnold L. Gesell: The Paradox of Nature and Nurture," 371.

237 "In replacing behavioral": Hulbert, *Raising America*, 173.

237 In *Infant and Child*: Arnold Gesell and Frances L. Ilg, *Infant and Child in the Culture of Today* (New York: Harper & Brothers, 1943). Quotations from, in order: 119, 144, 145, 325, 202.

237 The norms, he wrote: Ibid., 2.

237 He cautioned: Hulbert, *Raising America*, 166.

238 Norms should only: Gesell and Ilg, *Infant and Child*, 70.

238 Every mother would: J. Brennemann, "Pediatric Psychology and the Child Guidance Movement," *The Journal of Pediatrics* 2, no. 1 (1933), 20–21.

238 "The American craze": Grace Adams, *Your Child Is Normal: The Psychology of Early Childhood* (New York: Covici Friede, 1934). Quotations, in order: 13, 14, 19.

239 "There was a transformation": Thelen and Adolph, "Arnold L. Gesell: The Paradox of Nature and Nurture," 374.

241 "I Lie About": www.babble.com/toddler/toddler-development/toddler -children-physical-development-milestones (3 October 2012).

242 Across traditional: See Carol Delaney, "Making Babies in a Turkish Village," in *A World of Babies: Imagined Childcare Guides for Seven Societies*, eds. J. S. DeLoache and A. Gottlieb (New York: Cambridge University Press, 2000); K. E. Adolph, L. Karasik, and C. S. Tamis-LeMonda, "Motor Skills," in *Handbook of Cultural Developmental Science*, ed. M. Bornstein (New York: Taylor & Francis, 2010); and Rogoff, *The Cultural Nature of Human Development*.

242 The anthropologist David: Personal communication; "Infant Carrying and Prewalking Locomotor Development: Proximate and Evolutionary Perspectives," presented at the Conference of American Association of Physical Anthropologists, 2009.

243 Diarrhea can be deadly: M. F. Zeitlin et al., "Developmental, Behavioural, and Environmental Risk Factors for Diarrhoea among Rural Bangladeshi Children of Less than Two Years," *Journal of Diarrhoeal Disease Research* 13, no. 2 (1995).

243 The Western prejudice: See (for pictures and more) Béatrice Fontanel and Claire D'Harcourt, *Babies: History, Art, and Folklore* (Abrams: New York, 1997).

243 In his 1690: Calvert, *Children in the House*, 60.

244 In an 1835: Mintz, *Huck's Raft*, 80.

244 This laissez-faire: K. E. Adolph, "Motor and Physical Development: Locomotion," in *Encyclopedia of Infant and Early Childhood Development* (2008).

244 In 1900: A. Trettien, "Creeping and Walking," *The American Journal of Psychology* 12, no. 1 (1900): 1, 34.

245 "Skipping this": www.parenting.com/article/do-babies-need-to-crawl (3 October 2012).

245 "What's so important": www.medcentral.org/main/Whatssoimportant- aboutcrawling.aspx (3 October 2012).

246 If this parent searches: M. H. Immordino-Yang and K. Fischer, "Brain Development," in *The Corsini Encyclopedia of Psychology*, vol. 1, eds. I. B. Weiner and W. E. Craighead (New York: Wiley, 2010).

247 The Yale psychologist: E. Zigler, "A Plea to End the Use of the Patterning Treatment for Retarded Children," *American Journal of Orthopsychiatry* 51, no. 3 (1981). Also see the AAP's statement, "The Treatment of Neurologically Impaired Children Using Patterning," *Pediatrics* 104, no. 5 (1999).

247 "Walking without crawling": www.community.thebump.com/cs/ks /forums/thread/47647826.aspx (3 October 2012).

248 A study found: P. Robson, "Prewalking Locomotor Movements and Their Use in Predicting Standing and Walking," *Child: Care, Health, and Development* 10, no. 5 (1984).

16. WHERE MOVEMENT COMES FROM

249 "It seemed as": E. Thelen, "Motor Development: A New Synthesis," *American Psychologist* 50, no. 2 (1995): 79.

250 Esther Thelen was: Details of Thelen's early biography come from her charming essay "The Improvising Infant: Learning About Learning to Move," in *The Developmental Psychologists*, 28.

250 She was intrigued: Ibid., 34

251 Then she went out: Ibid., 35

252 It's so basic: L. Smith, "Movement Matters: The Contributions of Esther Thelen," *Biological Theory* 1, no. 1 (2006): 88.

252 The answer to Thelen's: Ibid.

254 As the psychologist Eugene: E. C. Goldfield, B. A. Kay, and W. H. Warren, "Infant Bouncing: The Assembly and Tuning of Action Systems," *Child Development* 64, no. 4 (1993): 1128.

255 Let us pause: B. Vereijken, "The Complexity of Childhood Development: Variability in Perspective," *Physical Therapy* 90, no. 12 (2010): 1854–55.

256 Being able to adjust: See K. Adolph, "Learning to Move," *Current Directions in Psychological Science* 17, no. 3 (2008).

257 What we assume: Daniela Corbetta, "Invited Commentary," *Physical Therapy* 89, no. 3 (2009).

257 In infancy, waking: K. E. Adolph and A. S. Joh, "Multiple Learning Mechanisms in the Development of Action," in *Learning and the Infant Mind*, eds. A. Woodward and A. Needham (New York: Oxford University Press, 2009).

257 An analysis of infants: Vereijken, "The Complexity of Childhood Development."

257 Another study found: S. E. Groen et al., "General Movements in Early Infancy Predict Neuromotor Development at 9 to 12 Years of Age," *Developmental Medicine and Child Neurology* 47, no. 11 (2005).

258 What seems clear: For more on this idea, see Mark S. Blumberg's *Freaks of Nature: What Anomalies Tell Us About Development and Evolution* (New York: Oxford University Press, 2009).

258 The average toddler: K. E. Adolph et al., "How Do You Learn to Walk?

Thousands of Steps and Dozens of Falls Per Day," *Psychological Science* 23, no. 11 (2012).

258 These reasons begin: See K. E. Adolph, "Learning to Keep Balance," in *Advances in Child Development & Behavior*, vol. 30, ed. R. Kail (Amsterdam: Elsevier Science, 2002).

259 To break the deadlock: W. Snapp-Childs and D. Corbetta, "Evidence of Early Strategies in Learning to Walk," *Infancy* 14, no. 1 (2009).

260 To do this, over: K. E. Adolph, B. Vereijken, and P. E. Shrout, "What Changes in Infant Walking and Why," *Child Development* 74, no. 2 (2003): 475.

261 As it happens: Adolph et al., "How Do You Learn to Walk?"

262 For a few centuries: Diana Dick, *Yesterday's Babies: A History of Babycare* (London: Bodley Head, 1987), 70.

263 In *The Principles*: William James, *The Principles of Psychology* (1890; reprint, New York: Dover, 1950), 406–407.

263 Instead of age: Adolph, "Learning to Move."

264 It's not that they: A. S. Joh and K. E. Adolph, "Learning from Falling," *Child Development* 77, no. 1: 94 (2006).

264 The average infant falls: Adolph and Joh, "Multiple Learning Mechanisms in the Development of Action."

265 Infants learn very little: Joh and Adolph, "Learning from Falling," 100.

265 In a follow-up study: K. E. Adolph et al., "Flexibility in the Development of Action," in *The Psychology of Action*, vol. 2, eds. J. Bargh, P. Gollwitzer, and E. Morsella (New York: Oxford University Press, 2008), 413.

265 They did the "hunchback": Ibid.

266 At nine months, most: Ibid., 419.

267 Take cruising: K. E. Adolph, S. E. Berger, and A. Leo, "Developmental Continuity? Crawling, Cruising, and Walking," *Developmental Science* 14, no. 2 (2011).

267 Which is why with: Adolph, "Learning to Keep Balance."

269 The vast majority: L. Sices, "Use of Developmental Milestones in Pediatric Residency Training and Practice: Time to Rethink the Meaning of the Mean," *Journal of Developmental and Behavioral Pediatrics* 28, no. 1 (2007).

269 "In other words": Ibid., 49

269 In a recent study: J. Darrah et al., "Intra-Individual Stability of Rate of Gross Motor Development in Full-Term Infants," *Early Human Development* 52 (1998).

271 A recent University of Virginia: D. Grissmer et al., "Fine Motor Skills and Early Comprehension of the World: Two New School Readiness Indicators," *Developmental Psychology* 46, no. 5 (2010).

271 Studies relying on: C. L. Ridgway, "Birth Size, Infant Weight Gain, and Motor Development Influence Adult Physical Performance," *Medicine and Science in Sports and Exercise* 41, no. 6 (2009).

271 Other studies, using: A. Taanila, "Infant Developmental Milestones: A 31-Year Follow-Up," *Developmental Medicine & Child Neurology* 47, no. 9 (2005); and G. K. Murray, "Infant Developmental Milestones and Subsequent Cognitive Function," *Annals of Neurology* 62, no. 2 (2007).

271 If you think I'm: B. A. Shaywitz, E. Ferrer, and S. E. Shaywitz, "A Case of Less Than Meets the Eye," *Annals of Neurology* 62, no. 2 (2007): 109.

17. JUST YOUR NORMAL MILESTONE-MEETING, SPEAR-THROWING INFANT

276 In the early 1970s: C. M. Super, "Environmental Effects on Motor Development: The Case of 'African Infant Precocity,'" *Developmental Medicine and Child Neurology* 18, no. 5 (1976).

276 "There is no": Ibid., 562.

277 Take bathing: Adolph, "Motor Skills," 67.

278 The first is taken: J. Mei, "The Northern Chinese Custom of Rearing Babies in Sandbags: Implications for Motor and Intellectual Development," in *Motor Development: Aspects of Normal and Delayed Development*, eds. J. H. A. van Rossum and J. I. Laszlo (Amsterdam: V. U. Uitgeverij, 1994).

279 The anthropologist Barry: Barry S. Hewlett, *Intimate Fathers: The Nature and Context of Aka Pygmy Paternal Infant Care* (Ann Arbor: University of Michigan Press, 1991), 32.

280 And Aka infants aren't: Rogoff, *The Cultural Nature of Human Development*, 5.

281 "There seems to be": B. Hopkins and T. Westra, "Maternal Expectations of Their Infants' Development: Some Cultural Differences," *Developmental Medicine and Child Neurology* 31, no. 3 (1989): 388.

281 The Kung, for example: A. Takada, "Motor-Infant Interactions among the !Xun," in *Hunter-Gatherer Childhoods*.

282 "Lifting the baby": H. Keller, R. D. Yovsi, and S. Voelker, "The Role of Motor Stimulation in Parental Ethnotheories: The Case of Cameroonian Nso and German Women," *Journal of Cross-Cultural Psychology* 33, no. 4 (2002): 408

282 "The Nso even": Heidi Keller, *Cultures of Infancy* (Mahwah, NJ: Lawrence Erlbaum, 2007), 119.

282 Their verdict was: Ibid., 121.

283 "Although they acknowledge": Keller, Yovsi, and Voelker, "The Role of Motor Stimulation in Parental Ethnotheories," 410.

18. THE MURKY ORIGINS OF BIPEDALISM, OR THE FIRST TODDLER

286 "If not for the realigned": Craig Stanford, *Upright: The Evolutionary Key to Becoming Human* (Boston: Houghton Mifflin, 2003), 49. Also see Daniel Lieberman's "Four Legs Good, Two Legs Fortuitous: Brains, Brawn, and

the Evolution of Human Bipedalism," in *In the Light of Evolution*, ed. J. B. Losos (Greenwood Village, CO: Roberts and Company, 2010).

287 In Tanzania: Stanford, *Upright*, 115–16.

288 The upright stance has: Lieberman, "Four Legs Good, Two Legs Fortuitous," 48.

288 The early anthropologist: Ales Hrdlicka, *Children Who Run on All Fours: And Other Animal-Like Behaviors in the Human Child* (New York: McGraw-Hill, 1931). Quotations: 249, 143, 192.

289 The basic cause: A. Hrdlicka, "Children Running Around on All Fours," *American Journal of Physical Anthropology* 11, no. 2 (1928): 30.

289 The siblings were: N. Humphrey, J. R. Skoyles, and R. Keynes, "Human Hand-Walkers: Five Siblings Who Never Stood Up," LSE Research Online, eprints.lse.ac.uk/463/ (3 October 2012).

289 the Turkish scientists: U. Tan, "A New Syndrome with Quadrupedal Gait, Primitive Speech, and Severe Mental Retardation as a Live Model for Human Evolution," *International Journal of Neuroscience* 116, no. 3 (2006).

290 Despite its constraints: see Blumberg, *Freaks of Nature*.

292 Kinesiologists will tell you: Adolph, Vereijken, and Shrout, "What Changes in Infant Walking and Why."

293 But mothers change: J. J. Campos, "Travel Broadens the Mind," *Infancy* 1, no. 2 (2000): 158–59.

293 A study conducted: L. B. Karasik, C. S. Tamis-LeMonda, and K. E. Adolph, "Transition from Crawling to Walking and Infants' Actions with Objects and People," *Child Development* 82, no. 4 (2011).

CODA, OR A BEDTIME STORY

295 "No other animal": William Kessen, "The American Child and Other Cultural Inventions," *American Psychologist* 34, no. 10 (1979): 815.

295 "'How do we know'": Selma Fraiberg, *The Magic Years: Understanding and Handling the Problems of Early Childhood* (New York: Scribner, 1959), 63.

298 All this is why: Carolyn Steedman, *Strange Dislocations: Childhood and the Idea of Human Interiority, 1780–1930* (London: Virago, 1995), 6.

299 Reviewing the book: Pamela Paul, "Kid Stuff," *New York Times*, October 1, 2009.

301 Jones found: For more on the debate over infant imitation, see Mark Blumberg's *Basic Instinct: The Genesis of Behavior* (New York: Basic Books, 2006).

302 The absence was so: Keller, *Cultures of Infancy*, 127.

303 Their mothers "discourage": Rogoff, *The Cultural Nature of Human Development*, 143.

303 A study comparing native: Ibid.

303 This dynamic holds: H. Keller, "Development as the Interface Between Biology and Culture: A Conceptualization of Early Ontogenetic Experiences," in *Between Culture and Biology: Perspectives on Ontogenetic*

Development, ed. H. Keller et al. (New York: Cambridge University Press, 2002).

304 Middle-class European: Rogoff, *The Cultural Nature of Human Development.*

304 You can see: H. Keller et al., "Developing Patterns of Parenting in Two Cultural Communities," *International Journal of Behavioral Development* 35, no. 3 (2011).

306 And yet, as Konner: Melvin Konner, *Childhood* (Boston: Little, Brown, 1991), 104.

306 The popular English: H. Hendrick, "Constructions and Reconstructions of British Childhood: An Interpretative Survey, 1800 to the Present," in *Constructing and Reconstructing Childhood: Contemporary Issues in the Sociological Study of Childhood,* eds. A. James and A. Prout (London: Falmer Press, 1997), 39.

307 "The Puritans were": S. Mintz, "The Evolution of Childhood: From the Puritans to the DVD Generation, A Look at Children in America," *Conscience,* Autumn 2006.

307 "If we are looking": Hugh Cunningham, *The Invention of Childhood* (London: BBC Books, 2006), 68.

307 "We have encountered": Ibid., 244.

Acknowledgments

Mark Oppenheimer, Abigail Dean, Andrew Iliff. Melissa Flashman, Nichole Argyres, Laura Chasen. Karen Adolph and Daniel Messinger. Blue State on Wall and Z&H on 57th. And the financial meltdown, which forced my hand.

With love and squalor: my parents, my sisters, Plainfield, Massachusetts, and its residents. Mila, who has a cameo at the end. Isaiah, who said when the book was finished: "Now I can learn about *babies!*"

Anya, without whom—well, not this. *Definitely* not this. Not much else either.

Index